ANIMAL CELLS

culture and media

ANIMAL CELLS
culture and media
ESSENTIAL DATA

D.C. Darling

Department of Molecular Medicine, The Rayne Institute, King's College School of Medicine, London, UK

S.J. Morgan

Leukaemia Research Fund, London, UK

JOHN WILEY & SONS
Chichester · New York · Brisbane · Toronto · Singapore

Published in association with BIOS Scientific Publishers Limited

©BIOS Scientific Publishers Limited, 1994. Published by John Wiley & Sons Ltd, Baffins Lane, Chichester, West Sussex PO19 1UD, UK, in association with BIOS Scientific Publishers Ltd, St Thomas House, Becket Street, Oxford OX1 1SJ, UK.

British Library Cataloguing in Publication Data
A catalogue record for this book is available from the British Library.

ISBN 0 471 94300 2

This book is dedicated to
Nicholas Morgan Darling

Library of Congress Cataloging in Publication Data
Darling, David C.
    Animal Cells: culture and media/D.C. Darling, S.J. Morgan.
        p. cm.—(Essential data series)
    Includes bibliographical references and index.
    ISBN 0-471-94300-2: $12.95
    1. Cell culture. 2. Animal cell biotechnology. I. Morgan, Sara J. II. Title. III. Series.
    QH585.2.D37 1994   94-31441
    591′.0724— —dc20   CIP

Typeset by Marksbury Typesetting Ltd, Bath, UK
Printed and bound in UK by H. Charlesworth & Co. Ltd, Huddersfield, UK

CONTENTS

Contents

Contents

ABBREVIATIONS

ACD	acid–citrate–dextrose
ADCC	antibody dependent cell cytotoxicity
aFGF	fibroblast growth factor (acidic)
AGIF	adipogenesis inhibitory factor
ATCC	American Type Culture Collection
BCDF	B-cell differentiation factor
BCGF	B-cell growth factor
BES	N,N'-bis(2-hydroxyethyl)-2-aminoethanesulfonic acid
bFGF	fibroblast growth factor (basic)
BHK	baby hamster kidney
BME	basal medium (Eagle's)
BMEE	basal medium Eagle's with Earle's
BMP	bone morphogenesis proteins
BPE	bovine pituitary extract
BSF	B-cell stimulatory factor

EBNA	Epstein–Barr nuclear antigen
EBSS	Earle's balanced salt solution
EBV	Epstein–Barr virus
ECACC	European Collection of Animal Cell Cultures
ECGF	endothelial cell growth factor
ECGS	endothelial cell growth supplement
ECM	extracellular matrix
EDF	eosinophil differentiation factor
EDTA	ethylenediaminetetraacetic acid
EGF	epidermal growth factor
EGTA	ethylene glycol-bis(β-aminoethyl ether) N,N,N',N'-tetraacetic acid
EHS	Engelbreth–Holm–Swarm
EMEM	Eagle's minimum essential medium
EMS	ethyl methanesulfonate
Epo	erythropoietin

BSS	balanced salt solution		FCS	fetal calf serum
BVD	bovine viral diarrhea		a-FGF	fibroblast growth factor (acidic)
CHO	Chinese hamster ovary		b-FGF	fibroblast growth factor (basic)
CIF	cartilage-inducing factor		G-CSF	granulocyte colony-stimulating factor
CLMF	cytotoxic lymphocyte maturation factor		GM-CSF	granulocyte–macrophage colony-stimulating factor
CMRL	Connaught Medical Research Laboratories			
CNDF	cholinergic neuronal differentiation factor		G-MEM	Glasgow minimal essential medium
ConA	concanavalin A		GRO/MGSA	GRO α/melanoma growth stimulating activity
CSF-2	colony stimulating factor			
CTL	cytotoxic T lymphocyte		HAT	hypoxanthine, aminopterin, thymidine
CNTF	ciliary neurotrophic factor		HBSS	Hanks' balanced salt solution
DIA	differentiation-inhibiting activity		HGF	hepatocyte growth factor
DIF	differentiation-inducing factor		HGPRT	hypoxanthine/guanine phosphoribosyl transferase
DMEM	Dulbecco's modification of Eagle's medium			
DMS	dimethyl sulfate		HILDA	human interleukin for DA cells
DMSO	dimethylsulfoxide		HIV	human immunodeficiency virus
DRF	differentiation retarding factor		HS	horse serum
DSF	decidual suppressor factor		HSF	hepatocyte-stimulating factor
dsRNA	double-stranded RNA		HSV-TK	herpes simplex virus thymidine kinase

IBR	infectious bovine rhinotracheitis
IFN	interferon
IgA-EF	IgA enhancing factor
IGF	insulin-like growth factor
IHS	heat-inactivated horse serum
IL	interleukin
IMEM	Iscove's modified Dulbecco's medium
IMP	inosine monophosphate
KGF	keratinocyte growth factor
KL	c-*kit* ligand
KRH	Krebs/Ringer/Hepes
L-15	Leibovitz's L-15 medium
LAF	lymphocyte activating factor
LAK	lymphokine-activated killer cells
LCLs	lymphoblastoid cell lines
LDL	low-density lipoprotein
LIF	leukemia inhibitory factor
LP	lymphopoietin
LPS	lipopolysaccharide
NAF	neutrophil-activating factor
NAP	neutrophil-activating peptide
NCS	newborn calf serum
NCTC	National Cancer Institute, Tissue Culture Section
NEAA	nonessential amino acids
N.F.	National Formulary
NGF	nerve growth factor
NK	natural killer
NKSF	natural killer cell stimulatory factor
OAF	osteoclast-activating factor
ORF	open reading frame
OSM	oncostatin M
PBS	phosphate-buffered saline
PDGF	platelet-derived growth factor
PEG	polyethylene glycol
PHA	phytohemagglutinin
PMA	phorbol myristate acetate
PMSF	phenyl methyl sulfonyl fluoride

LT	lymphotoxin	RANTES	regulated on activation, normal T cell expressed and secreted
LYNAP	lymphocyte-derived neutrophil-activating peptide	RCR	replication competent retrovirus
MCP	monocyte chemotactic protein	RPMI	Rosewell Park Memorial Institute
M-CSF	macrophage colony-stimulating factor	SCF	stem cell factor
MDNAP	monocyte-derived neutrophil-activating peptide	SDS	sodium dodecyl sulfate (sodium lauryl sulfate)
MDNCF	monocyte-derived neutrophil chemotactic factor	SLF	Steel cell factor
MEA	mast cell enhancing activity	SRBC	sheep red blood cells
MEM	minimal essential medium	STI	soyabean trypsin inhibitor
MGF	mast cell growth factor	TCGF	T-cell growth factor
MHC	major histocompatibility complex	TdT	terminal deoxynucleotide transferase
MIP	macrophage inflammatory protein	6-TG	6-thioguanine
MLPLI	melanoma-derived lipoprotein lipase inhibitor	TGF	transforming growth factor
MNNG	N-methyl-N'-nitro-N-nitrosoguanidine	TIF	tumor-inducing factor
MSA	multiplication stimulating activity	TK	thymidine kinase
MSV	Moloney murine sarcoma virus	TNF	tumor necrosis factor
		TRF	T-cell replacing factor

Abbreviations

TSH	thyroid stimulating hormone	w/v	weight/volume
USP	United States Pharmacopoeia	XMP	xanthine monophosphate
v/v	volume/volume		

PREFACE

Cell culture is an ever-widening field, which allows us to examine a basic unit of life under the inverted microscope. Once we have succeeded in keeping cells alive for long periods of time we can then start to manipulate them in order to ask further questions. To ensure that the answers obtained have any validity, then the conditions under which cells are routinely cultured must be the optimum for that type of cell. *Animal Cells: culture and media* seeks to enable the researcher to have the most relevant data available in a concise form for culture techniques to be performed. The aim is not to replace the excellent basic texts on cell culture that are available, but to complement them by being a ready reference that can fit into a lab coat pocket. We apologize for any missing data but hope that that which we have included will be of use.

D.C. Darling and S.J. Morgan

Chapter 1 **INTRODUCTION**

This Essential Data book is a pocket-sized ready reference for the bench worker in cell culture. In drawing together the data for this book we have utilized many sources of information and condensed them as much as possible into a compact, easy to read, tabular format. Although not intended as a techniques book, essential buffers and supplements are described where necessary and referenced where applicable.

Sections of the book include culture media, supplements, culture vessels, cell types, culture methods and safety. Culture media introduces choice, constituents, buffering systems and background to present-day media and balanced salt solution formulations, culminating in the essential requirements for making media.

A large supplements section covers the range of available 'additives' including serum, serum replacements and growth factors. A large range of cell types, both primary and continuous, are summarized with important relevant information.

Strategies for cell culture are also reviewed including primary cell extraction, day-to-day evaluation of cell lines, cryopreservation, cloning, modification of cells in culture and troubleshooting.

The final chapter gives manufacturers' addresses and a short appendix provides safety information

Chapter 2 **CULTURE MEDIA**

All media consist of an isotonic, buffered, basal nutrient medium which provides an energy source, coupled with inorganic salts, amino acids, vitamins and various supplements. These supplements, together with the difference in the mixture and concentration of the basic constituents, characterize each type of medium.

1 Choice of media

The type of medium recommended usually depends on the type of cell in culture (see Chapter 5). There are two main types: the simpler, more common types of medium (e.g. minimal essential medium (MEM), Rosewell Park Memorial Institute (RPMI) 1640 or Dulbecco's modification of Eagle's medium (DMEM)) and more complex/enriched media (e.g. Ham's F12, Iscove's modification of DMEM and Connaught

Glutamine
A most important constituent of any medium, now recognized as a major energy source along with glucose (a carbohydrate), and pyruvate (sometimes added). Exhaustion of glutamine has been shown to cause both reduced cell viability and productivity [1]. Low concentrations of other amino acids can also limit the maximum cell concentration and growth rate.

2.2 Vitamins

Most commonly the B-group vitamins are added (folic acid, biotin and pantothenate). Additionally, ascorbic acid (vitamin C) and α-tocopherol (vitamin E) are included in some media. Serum is an important source of vitamins

Medical Research Laboratories (CMRL) 1066). These complex media are used for more specialized cell types and also as the basis for the 'serum-free' media formulations. These are often used in an attempt to eliminate spurious results from the addition of unknown or ill-defined factors in serum. These factors present two main problems:

1. Substituting serum is difficult;
2. The relative importance of each component for each cell type is often not defined.

2 Media constituents

2.1 Amino acids

There are two types, 'essential' (*Table 1*) and 'nonessential' (*Table 2*). Essential amino acids are those not manufactured by the cell, plus cysteine and tyrosine. Individual requirements vary for the cell type being cultured. Nonessential amino acids are often added to medium to compensate for a particular cell type which is unable to manufacture them, or if they leach rapidly into the medium.

within a medium, when serum is reduced or eliminated an increased or novel requirement may become apparent. Low vitamin levels can affect cell survival and growth rate as their absence becomes limiting to metabolism (they act as cofactors).

2.3 Ions (constituents of media salts)

These are the major contributors to the osmolality of the medium and are mainly: Na^+, K^+, Mg^{2+}, Ca^{2+}, Cl^-, SO_4^{2-}, PO_4^{3-}, HCO_3^-.

Ca^{2+} and Mg^{2+} should be reduced for suspension cultures, as they are cofactors for attachment and cell aggregation. A medium designed for suspension cells, such as RPMI 1640, has a reduced level of these two salts. HCO_3^- levels are determined by the concentration of CO_2 in the incubator (i.e. in contact with the growth medium). $NaHCO_3/CO_2$ buffering is probably the most popular system used and requires a CO_2 level of 5–10% (recommendation dependent on the media used), and 100% humidity. This is known as an open system and the $NaHCO_3$ interacts with the medium as follows:

(1) $\quad H_2O + CO_2 \rightleftharpoons H_2CO_3 \rightleftharpoons H^+ + HCO_3^-$

(2) $\quad NaHCO_3 \rightleftharpoons Na^+ + HCO_3^-$

The products of cell metabolism (mostly CO_2), and the CO_2 in the incubator atmosphere interact with the water in the medium as shown in Equation 1. Thus the H^+ concentration is related to the CO_2 in the atmosphere. The $NaHCO_3$ in bicarbonate-buffered medium dissociates as detailed in Equation 2. These reactions are in a reversible equilibrium, and the system as a whole will tend to resist change in the ratio between the component parts. When the atmospheric concentration of CO_2 is regulated any increase in CO_2 and acidity (H^+), is prevented by a high HCO_3^- level achieved by the addition of $NaHCO_3$ (Equation 2). Interestingly, a secondary advantage of the use of $NaHCO_3$ (apart from the cost) is that the absence

3 Serum-free media

Serum-free media are based upon complex media which contain more amino acids, vitamins and other metabolites, such as nucleosides and lipids. The reduction or elimination of serum from a medium requires that ultra-pure reagents and water are used, as serum appears to contain protective, detoxifying proteins.

Serum-free media tend to be more cell-type specific, and therefore it is difficult to give a general guide to the types of serum-free media available. However, some have found wide usage, such as Iscove's modified DMEM and Ham's F12/DMEM mixture (1:1). Commercial suppliers provide both serum-free and reduced-serum media, for example AIM V media from Gibco, which is formulated for lymphocytes, and keratinocyte serum-free media. Serum-free media often need to be supplemented individually, depending on the cell type being cultured, therefore the literature for the particular cell type should be examined for further information on specific supplements, see also Chapter 3. Commercial

of either HCO_3^- or CO_2 appears to be limiting to cell growth [2].

2.4 Carbohydrate

Glucose is an energy source (along with glutamine) for most media. It is metabolized by glycolysis to form pyruvate which is then fed into the Krebs citric acid cycle via lactate or acetoacetate. In some cell types the Krebs cycle may not function entirely as normal and therefore their dependence on the alternative energy source, glutamine, may be very high. It is possible to substitute some of the glucose with an alternative carbohydrate such as galactose, fructose or mannose, this can reduce the build-up of lactic acid which acidifies the medium.

2.5 Other constituents

These include nucleosides, pyruvate and lipids which are added in more complex media. It is known that they are necessary when the serum level is reduced and can help during cloning and support of certain specialized cell types.

suppliers can provide serum-replacement mixtures. Cell growth in serum-free media is often slower than that found with serum-supplemented media. The main supplements required are insulin, transferrin and selenium, these seem to be universal requirements, other additions (e.g. bovine pituitary extract) can be as relatively undefined as the original serum.

Table 3 describes some of the most commonly used media and examples of their typical use.

4 Common media formulations

Media formulations can be found in any catalog that supplies media. *Table 4* lists the most popular basic formula suggested for suspension cells (RPMI 1640) and adherent cells (EMEM), and *Table 5* gives two examples of more complex types, Medium 199 and Ham's F12. They are listed side by side as mg l^{-1} and mM or μM for ease of comparison.

Culture Media

The formulations described here are specific, although variations are now available on the basic theme. Remember that the precise components may vary if the media is purchased either as a liquid or solid. L-Glutamine can be left out of liquid media (to improve its shelf life or to avoid its precipitation in $10 \times$ media), and more often included in powdered medium, whereas $NaHCO_3$ is most often omitted from powdered formulations. Additionally, some components may appear to vary between the liquid and solid media while in reality this is merely the anhydrous or the hydrated version (e.g. $CaCl_2$ or $CaCl_2.2H_2O$). These changes are often for convenience of preparation, or in some cases it may be that a crystalline component may settle out in a powder formulation.

Decisions on media are more difficult than they used to be. Novel formulations and new additions are available for each medium on a regular basis. For instance DMEM can be purchased in the high (4500 mg l^{-1}) or low (1000 mg l^{-1}) glucose concentration, both with and without sodium pyruvate, 110 mg l^{-1}. Hepes (5958 mg l^{-1}) modified

6 Essentials for making up media

Media bought from suppliers as a $1 \times$ solution is ready for the addition of supplements (see *Table 12*). A few substances may or may not have been already added to a $1 \times$ medium. For instance L-glutamine can be included in a $1 \times$ liquid, or a powdered formulation, but is very rarely included in a $10 \times$ formulation. Thus the strict definition of supplement will depend upon your starting point. Many workers make up media from powder or $10 \times$ liquid concentrate, and some of the more common media types can now be purchased as 'autoclavable' (i.e. bought as a powder, then partially made up and autoclaved prior to adding bicarbonate and other supplements as required). In this case the supplements are again different, as the heat-labile components are not included in the formulation, and must be added afterwards.

Some balanced salt solutions and buffers can also be bought as powders or $10 \times$ liquid concentrate.

The purchase of powdered or $10 \times$ liquid concentrate can

DMEM can now be found with or without sodium bicarbonate, with the osmotic balance maintained by a reduction in sodium chloride (from 6400 mg l^{-1} NaCl to 3500 mg ml^{-1} NaCl if 3700 mg l^{-1} NaHCO$_3$ retained). This is an example of just part of the choice available from just one supplier.

5 Balanced salt solutions

Balanced salt solutions are used to maintain cells for a short time in a viable condition rather than to promote their growth. They are often used for short incubations or for washing cells by centrifugation. During this time the cells need to be maintained in a medium of physiological pH and osmotic balance. The original solution used for this purpose was Ringer's, from which the solutions available today have evolved (*Table 6*). The formulations of some of these solutions are given in *Tables 7* and *8*. *Table 9* lists other common cell culture buffers, with their formulations being given in *Tables 10* and *11*.

reduce the cost of the materials, or can be more convenient if a lot is used and there are bulk requirements.

Table 12 gives the requirements for making medium from a 10 × liquid concentrate, and *Table 13* those for medium preparation from powder.

6.1 Sodium bicarbonate
NaHCO$_3$; molecular weight 84.

Usually made up as a 7.5% solution (7.5 g 100 ml^{-1}, 89 mM), sterilized through a 0.2 μm syringe filter. Can be stored in full aliquot for up to 3 years.

Remember that NaHCO$_3$ will dissociate in water as detailed in Section 2.3, and thus exchange with the immediate atmosphere. Thus, with a loose top, over a long period of time the concentration will drop.

DO NOT AUTOCLAVE.

6.2 Hepes

Many media formulations offer the additional option of Hepes (*N*-2-hydroxyethylpiperazine-*N'*-ethanesulfonic acid; molecular weight of free acid, 238.3, sodium salt, 260.3) buffered medium (10–25 mM). This should not replace the bicarbonate, as this is required by most mammalian cells [3]. In many cases the NaCl concentration is decreased in order to maintain the osmotic balance while maintaining the concentration of bicarbonate. Other sources recommend that, when combined, the concentration of either the Hepes or bicarbonate should not exceed 10 mM. Be aware of the possible osmotic consequences of the addition of Hepes to media not designed for this purpose.

6.3 L-Glutamine

$C_5H_{10}N_2O_3$; molecular weight 146.15.

Usually made up as a 200 mM solution (2.92 g 100 ml^{-1}), sterilized through a 0.2 µm syringe filter. Stored between claimed that it does not degrade with storage or incubation, this means that additional L-glutamine need not be added. Also, since there is no degradation, potentially harmful NH_3 build-up is prevented. Certainly (in our hands) cells appear to grow quite happily in the GlutaMAXI^{TM} medium supplement. GlutaMAXII^{TM} is an alternative containing the dipeptide glycyl-L-glutamine; whereas little or no adaption is required for GlutaMAXI^{TM}, this may not be the case for GlutaMAXII^{TM}.

6.4 Sodium hydroxide and hydrochloric acid

Sterile solutions of both 1 M NaOH and 1 M HCl can be prepared by filtration through 0.2 µm filters, though some filter membranes may not be compatible (check data sheet with the filter, there is conflicting advice as to what can be used, although some tests, i.e. 72 h at 25°C, seem a little rigorous for our purposes). Alternatively, sodium hydroxide made up under sterile conditions to 10 M, and then diluted into sterile water can be considered self-sterile. The same is true for HCl diluted from fresh 35% HCl.

−5°C and −20°C, stable for 18 months. L-Glutamine is an essential component for growth media: it is thus important to be aware of its stability in culture. Studies have shown [4] that at 4°C, 80% of L-glutamine remains after 3 weeks, whereas at 35°C a linear degradation was seen until, after 3 weeks, less than 15% of the original activity remained. This is in contrast to the stability seen with most other amino acids.

DO NOT AUTOCLAVE

Table 15 shows the amount of L-glutamine in various media types.

Recently Gibco-BRL have introduced an alternative to the use of glutamine, under the name of GlutaMAX™, which is a tradename covering a number of L-glutamine derivatives. GlutaMAX I™ is the dipeptide L-alanyl-L-glutamine. Used at the same mM concentration as is L-glutamine, it is

Sodium hydroxide can be autoclaved, but be sure to use autoclavable polypropylene, or good-quality glass to avoid leaching silicates from the glass.

Sodium hydroxide (NaOH; molecular weight 40), 10 M = 400 mg ml^{-1}; CARE, this is a very strong solution and dissolving is an exothermic reaction, a lot of heat is produced.

Point to note: 1 N (1 Normal solution) = 1 M solution. This is the case when a molecule can only dissociate to produce either one H$^+$ or one OH$^-$ ion, thus the same is true for HCl, but not, for instance, for H$_2$SO$_4$ where the molecule can dissociate to produce two H$^+$ ions.

Hydrochloric acid (HCl; molecular weight 36.46), 35% = 4.06 M solution. To make 1 M solution, 1 ml + 3.06 ml H$_2$O; CARE, when diluting acids always add the acid to the water and not the other way round.

Culture Media

Table 1. Essential amino acids

Amino acid	Formula weight
L-Arginine	174.2 (free base)
	210.7 (hydrochloride)
L-Cysteine	240.3 (free base)
	313.2 (dihydrochloride)
L-Glutamine	146.1
L-Histidine	155.2 (free base)
	209.6 (hydrochloride monohydrate)
L-Isoleucine	131.2
L-Leucine	131.2
L-Lysine	146.2 (free base)
	182.6 (hydrochloride)
L-Methionine	149.2
L-Phenylalanine	165.2
L-Threonine	119.1
L-Tryptophan	204.2
L-Tyrosine	181.2 (free base)
	225.2 (disodium salt)
	217.7 (hydrochloride)
L-Valine	117.1

Table 2. Nonessential amino acids[a]

Amino acid	Formula weight	1 × solution mg l^{-1}	mM
L-Alanine	89.09	8.9	0.1
L-Asparagine	132.1	15.0	0.1
	150.1 (hydrate)	—	—
L-Aspartic acid	133.1	13.3	0.1
L-Glutamic acid	147.1	14.7	0.1
	169.1 (Na salt)	—	—
Glycine	75.07	7.5	0.1
L-Proline	115.1	11.5	0.1
L-Serine	105.1	10.5	0.1

[a]MEM nonessential amino acids [5].

Table 3. Selection of some common media types

Media type	Brief description
BME	Basal medium (Eagle). Early medium used for mouse L cells and HeLa cells. This is often used as a washing buffer as the Hepes will hold the pH in the presence of atmospheric O_2 [6, 7]
MEM	Minimal essential medium (Eagle's), EMEM. Improvement of BME, with higher concentrations of amino acids, also contains Earle's salts for $CO_2/NaHCO_3$ buffering [5]
DMEM	Dulbecco's modification of Eagle's medium. Four times the amino acid and vitamin concentration of BME with additional amino acids and ferric nitrate. Glucose often now added to 4500 mg l^{-1} (either with or without sodium pyruvate), whereas originally set at 1000 mg l^{-1} with the addition of sodium pyruvate (110 mg l^{-1}). Originally used for embryonic mouse and virally infected hamster cells [8]
Iscoves	Iscove's modification of Dulbecco's medium. With additional amino acids and vitamins, also selenium, sodium pyruvate and Hepes. Potassium nitrate replaces ferric nitrate. Used for serum-free growth of lymphocytes and hybridomas [9]
199	Medium 199. Developed as a protein-free medium, although serum sometimes added. Based on Earle's balanced salt solution (EBSS) plus a large number of amino acids, vitamins, growth factors and lipids [10]
McCoy's 5A	Based on BME but with the addition of the amino acid + vitamin mixture from Medium 199. Originally used for primary culture of small biopsy specimens [11, 12]
RPMI 1640	Rosewell Park Memorial Institute 1640. For long-term culture of blood cells, a modification of McCoy's 5A medium now used as a general medium (with serum) for lymphocyte and hybridoma cultures [13]
CMRL 1066	Connaught Medical Research Laboratories 1066. A simplified modification of Medium 199. Originally used a serum-free formulation for murine L-cells [14]
Ham's F12	Developed for cloning hamster ovary cells. Originally used as serum-free, but now used with serum supplement for a variety of cell types. Used 1 : 1 with DMEM for a serum-free formulation and many primary culture applications [15]
MCDB series	Originating in the Department of Molecular, Cellular and Developmental Biology at the University of Colorado. A series of complex media formulations developed from Ham's F12 for growth of various nontransformed cell types in serum-free media. Each subcategory containing supplements optimized for a specific cell type. MCDB 153 – human epidermal keratinocyte culture; MCDB 104 – orginally for human diploid fibroblast culture, i.e. MRC-5 [16–18]

Culture Media

Table 4. Formulation of RPMI and EMEM

	RPMI1640		EMEM	
	mg ml^{-1}	mM	mg ml^{-1}	mM
Inorganic salts				
CaCl$_2$	—	—	200	1.8
Ca(NO$_3$)$_2$.4H$_2$O	100	0.42	—	—
KCl	400	5.4	400	5.4
MgSO$_4$.7H$_2$O	100	0.41	200	0.8
NaCl	6000	100	6800	117.0
NaHCO$_3$	2000	23.8	2200	26.19
NaH$_2$PO$_4$.2H$_2$O	—	—	150	0.96
Na$_2$HPO$_4$	800	5.63	—	—
Amino acids				
L-Arginine	200	1.15	—	—
L-Arginine.HCl	—	—	126.4	0.6
L-Asparagine	50	0.43	—	—
L-Aspartic acid	20	0.15	—	—
L-Cysteine	50	0.21	24.02	0.1
L-Glutamic acid	20	0.14	—	—
L-Glutamine	300	2.05	292.3	2
Glycine	10	0.13	—	—
L-Histidine.HCl.H$_2$O	—	—	41.9	0.2
L-Histidine	15	0.1	—	—

	RPMI1640 mg l^{-1}	RPMI1640 µM	EMEM mg l^{-1}	EMEM µM
L-Hydroxyproline	20	0.15	—	—
L-Isoleucine	50	0.38	52.5	0.4
L-Leucine	50	0.38	52.3	0.4
L-Lysine.HCl	40	0.27	73.06	0.4
L-Methionine	15	0.1	14.9	0.1
L-Phenylalanine	15	0.09	33.02	0.2
L-Proline	20	0.17	—	—
L-Serine	30	0.29	—	—
L-Threonine	20	0.17	47.64	0.4
L-Tryptophan	5	0.02	10.2	0.05
L-Tyrosine	20	0.11	36.22	0.2
L-Valine	20	0.17	46.9	0.4
Other components				
Glutathione	1	0.0326	—	—
D-Glucose	2000	11.1	1000	5.55
Phenol Red	5	0.0133	10	0.02655

	RPMI1640		EMEM	
	mg l^{-1}	µM	mg l^{-1}	µM
Vitamins/cofactors				—
Biotin	0.2	0.82		8.3
Choline chloride	3	21.4	1	2.3
Folic acid	1	2.27	1	11
i-Inositol	35	194.4	2	

Continued

Table 4. Formulation of RPMI and EMEM, *continued*

	RPMI1640		EMEM	
	mg l^{-1}	µM	mg l^{-1}	µM
Nicotinamide	1	8.2	1	8.2
p-Aminobenzoic acid	1	7.3	—	—
Pantothenate.Ca	0.25	0.52	1	4.6
Pyridoxal.HCl	—	—	1	6
Pyridoxine.HCl	1	4.9	—	—
Riboflavin	0.2	0.5	0.1	0.27
Thiamine.HCl	1	3.0	1	3
Vitamin B$_{12}$	0.005	0.0036	—	—

Points to note:

Observe that the calcium concentration is elevated in EMEM, this is used by many adherent cells as an attachment cofactor.

EMEM contains all the essential amino acids, as would be expected (*Table 1*), and none of those termed 'nonessential' (thus: minimal essential medium). RPMI 1640 includes all the nonessential amino acids (NEAA) with the exception of alanine (*Table 2*), thus in modifying the formula for one's own use there would be little to be gained by adding NEAAs to RPMI1640.

Sodium succinate is no longer added to some EMEM formulations, although some companies still include it as sodium succinate.6H$_2$O 100 mg l^{-1}.

Table 5. Formulation of Medium 199 and Ham's F12

	199		Ham's F12	
	mg ml^{-1}	mM	mg ml^{-1}	mM
Inorganic salts				
$CaCl_2.2H_2O$	264.9	1.8	44.1	0.3
KCl	400	5.4	223.7	3
$MgSO_4.7H_2O$	200	0.8	147.8	0.6
NaCl	6800	117	7600	131
$NaHCO_3$	2200	26.19	1176	14
$NaH_2PO_4.2H_2O$	158.3	1.01	—	—
Na_2HPO_4	—	—	142	1
Amino acids				
L-Alanine	25	0.28	8.91	0.1
L-Arginine.HCl	70	0.33	210.7	1
L-Asparagine.H_2O	—	—	15.01	0.1
L-Aspartic acid	30	0.23	13.31	0.1
L-Cysteine.HCl	0.1	6.3×10^{-4}	31.53	0.2
L-Cystine	20	0.083	—	—
L-Glutamic acid	66.82	0.45	14.71	0.1
L-Glutamine	100	0.68	146.2	1
Glycine	50	0.66	7.51	0.1
L-Histidine.HCl.H_2O	21.88	0.1	20.96	0.1
L-Hydroxyproline	10	0.076	—	—
L-Isoleucine	20	0.15	3.94	0.03
Continued				

Table 5. Formulation of Medium 199 and Ham's F12, *continued*

	199		Ham's F12	
	mg ml^{-1}	mM	mg ml^{-1}	mM
L-Leucine	60	0.46	13.12	0.1
L-Lysine.HCl	70	0.38	36.53	0.2
L-Methionine	15	0.1	4.48	0.03
L-Phenylalanine	25	0.15	4.96	0.03
L-Proline	40	0.35	34.54	0.03
L-Serine	25	0.24	10.51	0.1
L-Threonine	30	0.25	11.91	0.1
L-Tryptophan	10	0.049	2.04	0.01
L-Tyrosine	40	0.22	5.4	0.03
L-Valine	25	0.21	11.72	0.1
Other components				
Phenol Red	10	26.55	10	26.55
D-Glucose	1000	5.55	1802	10
Pyruvate.Na	—	—	110	1

	199		Ham's F12	
	mg l^{-1}	μM	mg l^{-1}	μM
Vitamins/cofactors				
Ascorbic acid	0.05	0.28	—	—
Biotin	0.01	0.04	0.0073	0.03
Calciferol	0.1	0.25	—	—

Choline chloride	0.5	3.57	13.96	100
Folic acid	0.01	0.02	1.32	3
i-Inositol	0.05	0.28	18.02	100
Menadione	0.01	0.06	—	—
Nicotinamide	0.025	0.21	0.037	0.3
Nicotinic acid	0.025	0.2	—	—
p-Aminobenzoic acid	0.05	0.37	—	—
Pantothenate.Ca	0.01	0.02	0.48	1
Pyridoxal.HCl	0.025	0.12	—	—
Pyridoxine.HCl	0.025	0.12	0.062	0.3
Riboflavin	0.01	0.02	0.038	0.1
Thiamine.HCl	0.01	0.03	0.337	1
α-Tocopherol.PO$_4$.2Na	0.01	0.02	—	—
Vitamin A acetate	0.01	0.33	—	—
Vitamin B$_{12}$	—	—	1.36	0.0037
Trace elements				
CuSO$_4$.5H$_2$O	—	—	2.5	0.01
FeSO$_4$.7H$_2$O	—	—	834	3
Fe(NO$_3$)$_3$.9H$_2$O	720	1.8	—	—
ZnSO$_4$.7H$_2$O	—	—	863	3
Additional components				
Adenine sulfate	10	27	—	—
ATP.2Na	1	1.81	—	—
5-AMP	0.2	0.58	—	—
Continued				

Culture Media

Table 5. Formulation of Medium 199 and Ham's F12, *continued*

	199		Ham's F12	
	mg l^{-1}	μM	mg l^{-1}	μM
Cholesterol	0.2	0.52	—	—
2-Deoxy D-ribose	0.5	3.73	—	—
Guanine.HCl	0.3	1.61	—	—
Glutathione	0.05	0.16	—	—
Hypoxanthine	0.3	2.21	4.08	30
Linoleate.Me	—	—	0.088	0.3
Lipoic acid	—	—	0.206	1
Putresine.2HCl	—	—	0.161	1
D-Ribose	0.5	3.31	—	—
Sodium acetate	50	0.61	—	—
Thymine	0.3	2.38	—	—
Thymidine	—	—	0.727	3
Tween 80	20	15.27	—	—
Uracil	0.3	2.68	—	—
Xanthine	0.3	1.97	—	—

Points to note:

Medium 199 and Ham's F12 contain a similar base amino acid composition as RPMI 1640, although the 199 medium appears to be the richer of the three, with the notable exception of the relatively huge asparagine content of Ham's F12.

Major differences in the more complex media are apparent in the additional vitamins/cofactors and the 'additional components'.

Table 6. Balanced salt solutions

Solution	Brief description
Ringer's	Developed from inorganic salts, calcium, potassium and sodium to approximate that found in sea water. Has been modified to give salt solutions of improved quality [19]
Tyrode's	Another early solution, although still in use today. Problems can occur with calcium precipitation, other solutions are preferable [20]
Earle's	Contains sodium bicarbonate and should be used with CO_2 (5%). It is the basis of Eagle's medium and hence other media developed from this. Quite widely used [21]
Hanks'	Widely used, designed for buffering in air rather than in a CO_2 atmosphere [22]
Gey's	Not widely available commercially, although it can be obtained as two alternative formulations, either with or without bicarbonate buffering [23]
Puck's A	Can be used either with or without bicarbonate, for use as a trypsin diluent. Can be used with CO_2 where HCO_3^- added. No Mg^{2+} or Ca^{2+} salts [24]
Puck's G	Contains Mg^{2+} and Ca^{2+} salts but no bicarbonate. For use in a non-CO_2 atmosphere [25]
Paul's	Tris-buffered salt solution, a more effective buffer than phosphate buffers (as above). Contains high level of glucose and also citric acid [26]
Eagle's spinner salts	Essentially the same as Earle's salts, but without $CaCl_2$ and reduced NaH_2PO_4. Low calcium wash buffer for use in spinner cultures [5]

Table 7. Balanced salt solution formulations: EBSS, HBSS and Tyrode's

	EBSS[a]		HBSS[b]		Tyrode's	
	mg l⁻¹	mM	mg l⁻¹	mM	mg l⁻¹	nM
Inorganic salts						
NaCl	6800	117	8000	139	8000	139
KCl	400	5.4	400	5.4	200	2.7
CaCl₂	200	1.5	140	1.1	200	1.5
MgCl₂	—	—	—	—	45.7	0.48
MgSO₄	96	0.8	96	0.8	—	—
Na₂HPO₄	—	—	47.7	0.34	—	—
NaH₂PO₄	108	0.9	—	—	38	0.32
KH₂PO₄	—	—	60	0.44	—	—
NaHCO₃	2200	26.16	350	4.16	1000	11.89
Other components						
Glucose	1000	5.56	1000	5.56	1000	5.56
Phenol Red	10	0.025	10[c]	0.025	—	—

[a]EBSS, Earle's balanced salt solution.

[b]HBSS, Hanks' balanced salt solution.

[c]Original formulation called for 20 mg l⁻¹ Phenol Red, though most companies now supply at 10 mg l⁻¹.

EBSS, containing bicarbonate, is recommended for CO₂ atmosphere buffering (or can be equilibrated against 5% CO₂ before use), whilst the HBSS is generally used for buffering in air.

Table 8. Balanced salt solution formulations: GBSS[a] and Puck's A and G

	GBSS[c]		Puck's A		Puck's G	
	mg l⁻¹	mM	mg l⁻¹	mM	mg l⁻¹	mM
Inorganic salts						
	(7000)	(121.6)				
NaCl	8000	139	8000	139	8000	139
KCl	370	4.96	400	5.4	400	5.4
CaCl₂	133.5	1.2	—	—	12	0.11
MgCl₂	98.4	1.03	—	—	—	—
MgSO₄	34	28	—	—	75	0.625
Na₂HPO₄	120	0.85	—	—	81.5	0.57
	(26.1)	(0.22)				
NaH₂PO₄	(—	—)	—	—	—	
KH₂PO₄	30	0.22	—	—	150	1.1
	(2270)	(26.9)				
NaHCO₃	227	2.69	350	4.16	—	—
Other components						
Glucose	1000	5.56	1000	5.56	1100	6.12
Phenol Red	—	—	5[b]	0.0125	1.2	0.003

[a]GBSS, Gey's balanced salt solution.

[b]Originally required 20 mg l⁻¹, mostly commercially available at 5 mg l⁻¹.

Figures superscripted in brackets represent alternative formulation with increased bicarbonate buffering with the molality maintained by reduced salt.

Puck's A and Puck's G, shown is the modification of Puck's A to Puck's G; the original formulation (A) representing bicarbonate-buffered saline to the more sophisticated (G) which omits the bicarbonate and includes calcium, magnesium and phosphates.

Table 9. Other common cell culture buffers

Dulbecco's	Phosphate-buffered saline (PBS). Can be used as a Ca^{2+}/Mg^{2+}-free solution for use in trypsin solutions. Also available with calcium and magnesium salts if required. A good general washing buffer where incubation is not necessary [26]
Alsever's	Citrate-based buffer, popular choice for the short-term storage of sheep red blood cells for use in rosetting [28]
BME	Basal medium Eagle's. Many versions available, formulation shown here replaces bicarbonate with 10 mM (can use up to 20 mM) Hepes, and incorporates Earle's salts. BME Earle's (BMEE) for use as a more sophisticated wash (and storage) buffer that buffers in air [6, 7]

Table 10. Dulbecco's PBS (DPBS) and Alsever's solution

	DPBS		Alsever's	
	mg l^{-1}	mM	mg l^{-1}	nM
NaCl	8000	139	4200	73
KCl	200	2.7	—	—
CaCl$_2$	100	0.75	—	—
MgCl$_2$	45.7	0.48	—	—
Na$_2$HPO$_4$	1150	8.1	—	—
KH$_2$PO$_4$	200	1.47	—	—
C$_6$H$_8$O$_7$[a]	—	—	550	2.6
C$_6$H$_5$O$_7$Na$_3$2H$_2$O[b]	—	—	8000	27
Glucose	—	—	20 500	114

[a]Citric acid.

[b]Trisodium citrate dihydrate.

DPBS is often used as a Ca^{2+}- and Mg^{2+}-free formulation for trypsin treatment. Omission of these salts has little effect on the osmotic balance (289±15 to 285±14 mOsm kg^{-1} H$_2$O).

Most cells remain happy provided that they are within the range 260–320 mOsm kg^{-1} H$_2$O.

Table 11. Formulation of basal medium Eagle with Earle's salts (BMEE)

	BMEE	
	mg l^{-1}	mM
Inorganic salts		
CaCl$_2$	200	1.8
KCl	400	5.4
MgSO$_4$.7H$_2$O	200	0.8
NaCl	6800	117
NaH$_2$PO$_4$.2H$_2$O	158.3	1
Amino acids		
L-Arginine.HCl	21.06	0.1
L-Cysteine	12	0.05
L-Glutamine	292.3	2
L-Histidine.HCl.H$_2$O	10.48	0.05
L-Isoleucine	26.23	0.2
L-Leucine	26.23	0.2
L-Lysine.HCl	36.53	0.2
L-Methionine	7.46	0.05
L-Phenylalanine	16.5	0.1
L-Threonine	23.82	0.2
L-Tryptophan	4.08	0.02
L-Tyrosine	18.11	0.1
L-Valine	23.43	0.2

Other components		
D-Glucose	1000	5.55
Phenol Red	10	0.02655
Hepes	2383	10

	mg l^{-1}	μM
Vitamins/cofactors		
Biotin	1	4.1
Choline chloride	1	8.3
Folic acid	1	2.27
i-Inositol	2	11
Nicotinamide	1	8.2
Pantothenate.Ca	1	4.6
Pyridoxal.HCl	1	6
Riboflavin	0.1	0.27
Thiamine.HCl	1	3.0

Culture Media

Table 12. Requirements for making medium from $10 \times$ liquid concentrate

850 ml sterile, pure water in suitable media bottle
Sterile, pure water (minimum double distilled; for topping up)
7.5% $NaHCO_3$, sterile liquid
1 M NaOH and 1 M HCl, sterile
Media concentrate, $10 \times$, as necessary

The amount of bicarbonate required is dependent upon the type of medium being prepared, this must be checked with the manufacturer's instructions (see *Table 14*). Once all items are added then the final volume can be completed with the spare sterile pure water.

Table 14. Bicarbonate additions for various media types

	$NaHCO_3$		
Medium type	ml l^{-1} of 7.5% solution	g l^{-1} of powder	mM
Ames	25.7	1.932	22.9
BGJb	46.6	3.5	41.6
BME with EBSS[a]	29.3	2.2	26
BME with HBSS[b]	4.7	0.35	4
CMRL-1066	29.3	2.2	26
CRCM-30	8.0	0.6	7.1
DMEM	49.3	3.7	44
DMEM/F12 1:1[c]	16.0	1.2	14.2
EBSS	29.3	2.2	26
Fischers	15.0	1.125	13.5
Gey's GBSS[d]	30.2[(3.02)]	2.27[(0.227)]	272.7[(27.27)]
Glasgow MEM (G-MEM)	36.7	2.75	33
Ham's F10	16.0	1.2	14.2
Ham's F12	15.7	1.176	14
HBSS	4.7	0.35	4
Iscove's (IMEM)	40.4	3.024	36
Jolik's MEM	26.6	2.0	23.8
McCoy's 5A (RPMI 1629)	29.3	2.2	26
MCDB-151	15.7	1.176	14

Table 13. Media preparation from powder – requirements

Sterile media bottles, enough for use
Supply of pure water for diluting powder
Sodium bicarbonate powder
Filtration system
Powdered medium, as necessary
1 M HCl or NaOH as appropriate
Spare sterile cap for last bottle filtered

The filtration system used should be as trouble-free as possible to minimize the risk of contamination at this vital stage of preparation of medium. A positive-pressure system tends to be the easiest type as it is rapid. Slow filtration through small, limited capacity 0.22 μm filters should be avoided.

It is not recommended that concentrated media be made up from powder. It is possible to filter many liters of medium in a short time using a capacity positive-pressure system; however, it should be remembered that it is preferable to limit the time that a ready-made medium is sitting on the shelf (even at low temperature).

MCDB-153	15.7	1.176	14
MCDB-302	15.7	1.176	14
Medium 199 with EBSS	29.3	2.2	26
Medium 199 with HBSS	4.7	0.35	4
MEM-α	26.6	2.0	23.8
MEM with HBSS	4.7	0.35	4
MEM with EBSS	29.3	2.2	26
NCTC 109	29.3	2.2	26
NCTC 135	29.3	2.2	26
Puck's A	4.7	0.35	4
RPMI 1640	26.7	2.0	23.8
RPMI Dutch mod	13.3	1.0	11.9
RPMI Searle	26.7	2.0	23.8
Tyrodes	13.3	1.0	11.9
Waymouth's MB 752/1	29.9	2.24	26.6
William's Medium E	29.3	2.2	26

[a]EBSS, Earle's balanced salt solution.
[b]HBSS, Hanks' balanced salt solution.
[c]DMEM/F12 1:1 – note that commercial preparations have a bicarbonate concentration not consistent with a 1:1 mixture. Thus those with home-made formulations will need to decide which bicarbonate concentration to follow.
[d]Figures in superscript represent a low bicarbonate variation.

Table 15. L-Glutamine additions for various media types

Medium type	L-Glutamine		
	ml l^{-1} of 200 mM solution	g l^{-1} of powder	mM
Ames	2.5	0.073	0.5
BGJb	6.85	0.2	1.37
BME with EBSS[a]	10.0	0.292	2
BME with HBSS[b]	10.0	0.292	2
CMRL-1066	3.4	0.1	0.68
CRCM-30	15.0	0.439	3
DMEM	20.0	0.584	4
DMEM/F12 1:1[c]	12.5	0.365	2.5
Fischers	6.85	0.2	1.37
Glasgow MEM (G-MEM)	20.0	0.584	4
Ham's F10	5.0	0.146	1
Ham's F12	5.0	0.146	1
Iscove's (IMEM)	20.0	0.584	4
Jolik's MEM	10.1	0.294	2.01
McCoy's 5A (RPMI 1629)	7.5	0.219	1.5
MCDB-105	12.5	0.365	2.5
MCDB-110	12.5	0.3653	2.5
MCDB-131	50	1.461	10

MCDB-151	30	0.8772	6
MCDB-153	30	0.8772	6
MCDB-201	5.0	0.1461	1
MCDB-302	15.0	0.4386	3
Medium 199 with EBSS	3.4	0.1	0.68
Medium 199 with HBSS	3.4	0.1	0.68
MEM-α	10.0	0.292	2
MEM with HBSS	10.0	0.292	2
MEM with EBSS	10.0	0.292	2
NCTC 109	4.6	0.136	0.93
NCTC 135	4.6	0.136	0.93
RPMI 1640	10.3	0.3	2.05
RPMI Dutch mod	10.3	0.3	2.05
RPMI Searle	10.3	0.3	2.05
Waymouth's MB 752/1	12.0	0.35	2.4
William's Medium E	10.0	0.292	2

[a]–[d], see footnotes [a]–[d] for *Table 14*.

Culture Media

Chapter 3 **SUPPLEMENTS TO MEDIA**

Supplements to media range from the addition of serum and antibiotics to serum replacements, attachment and growth factors for more demanding cell types. Additionally, this chapter will cover some of the formulations for serum-free media which are becoming more commonplace in cell-culture laboratories.

The lists in this chapter will give the most usual additives and their concentrations and, where necessary, the constituents.

1 Serum

Serum or serum-like replacements are necessary for the growth of most cell types. Serum itself contains an ill-defined mixture of substances, many of which are necessary for the proliferation of cells. This mixture includes proteins such as and test routinely. Serum must also be incubated for 30 min at 56°C to inactivate complement that may react with the globulins also present in the serum. Serum should be tested prior to purchase on the cell types to be cultured to ensure that the best possible serum is being obtained for the cell growth or experimentation. Once a batch of serum has been selected then it should be reserved to ensure a continuous supply of optimal serum.

There are several types of serum: fetal bovine, new-born calf, donor calf, calf, horse. These are the most common types available; however, the following are also available for use: human, human type AB, goat, chicken, rabbit, pig, mouse, rat, lamb and guinea-pig.

Of the more common sera, fetal bovine is the most

globulins, albumin, transferrin, fetuin and fibronectin. They are carriers for minerals or hormones, or can be hormones themselves. Polypeptides, such as platelet-derived growth factor (PDGF) and epidermal growth factor (EGF), are also present. Hormones (e.g. insulin and hydrocortisone) have marked effects on the growth of cells. Minerals and trace elements are present; various inhibitory substances can also be present. An attempt to replace serum needs to take into account all the possible factors that may affect proliferation of a particular cell type as all cell types may have different requirements. However, for some cell types the replacements have been defined in the literature. Most cell-culture suppliers now have serum replacement/serum-like substances which are still partially undefined but may be more consistent than serum itself. They usually have a special proprietary mixture of substances which are not disclosed.

To use serum it should be determined that it has been ultrafiltered and tested for various contaminants, particularly mycoplasma; most companies now extensively treat

expensive, and therefore sometimes it is possible to grow some cell types in new-born calf or horse serum to reduce costs. The types of cell lines that can be grown in this way are given in Chapter 5. The possible choices of fetal calf serum are listed in *Table 1*.

Sera with specialized applications, or for fastidious cells, should be batch tested (while holding a reserve) to check whether the sample is good for the cells under study. Some treatments may reduce the growth-promoting ability for some cell types and yet increase it for others, and, paradoxically, reducing the serum level, say from 10% to 5%, may improve growth. For instance, serum can contain both PDGF and transforming growth factor (TGF) which can promote the growth of fibroblasts (PDGF) while reducing the growth of epithelial cells (TGF). All sera will have been routinely sterile-filtered and screened for the presence of contaminants such as endotoxin, mycoplasm, virus (including, bovine viral diarrhea (BVD), bovine adenovirus, parvovirus, infectious bovine rhinotracheitis (IBR), etc.).

Supplements to Media

2 Serum replacements

Every cell line/cell type is different, but usually they are grown in 5–10% serum when a standard medium is being used. Should a more complex medium (those with 'serum-free' potential) be used, then the amount of serum can sometimes be reduced, and may be replaced.

Serum-replacement supplements can represent the first step on the road to a serum-free culture. However, in many cases these replacements are actually reduced serum coupled with additional, defined, growth factor additions, and can include up to 25% fetal or new-born calf serum.

3 Serum-free media

Many serum-free media have now been designed on a cell type by cell type basis. These media originated in the work of people like Barnes and Sato [1] and Macaig *et al.* [2] in the early 1980s. Since then a number of serum-free media formulations have come on the market (*Table 2*). These tend

essential metabolites and make them available to cells in a usable form and also remove potentially toxic metabolites. Bovine serum albumin and transferrin are transport factors which are normally found in serum, therefore when using serum-free media these need to be added.

5 Antibiotics

The use of antibiotics is widespread (*Table 3*) but controversial. It is routinely recommended that antibiotics should not be used in the long term because of the possibility of masking possible infections present in the culture. In addition, it is thought that the continuous use of antibiotics can breed certain antibiotic-resistant strains of contaminants within the cell culture laboratory, which are subsequently difficult to clear. Having stated these points it must be said that many laboratories do indeed use antibiotics on a regular basis and rely on occasional testing for contaminants. The antibiotics used in this way are most often penicillin, streptomycin and gentamicin.

to be sold on the basis of the cell type for which they have been designed, and some contain proprietary information, meaning that the additions may not be known.

4 Other media additives

Hormones are the nonnutritional factors required by some cell types. They are generally simpler than growth factors and can promote intracellular metabolism. One in particular, insulin, appears to be almost obligatory for any serum-free medium formulation. Others include estradiol, thyroxine, triiodothyronine, hydrocortisone (cortisol), dexamethasone, progesterone and glucagon. Hydrocortisone and dexamethasone are in the glucocorticoid family. Other steroid hormones are estradiol, testosterone and progesterone. Thyroid hormones are also required for some cell lines, usually triiodothyronine, also known as T_3.

Other media additives include transport factors, which bind

Some fungicides are particularly toxic to some cell types and care should be exercised when using these. They should not be included in cultures on a regular basis.

An additional hazard is the use of more than one antibiotic in the culture system, the combined effects can be potentially toxic.

6 Growth factors

Some media supplements are added for their growth-promoting quality rather than their nutritional value. They may seem to have mitogenic activity but their functions are not well understood. Growth factors are used to supplement serum-free medium and culture conditions with reduced serum; also for more specialized cell lines and for primary cell culture. *Table 4* gives the recommended concentrations for use of the most common factors (growth/cytokine/hormone), their application and the different types currently available. *Table 5* lists the shortened and alternative names of these factors.

Table 1.   Choices of fetal calf serum

Type	Applications
Basic fetal calf serum	Many – basic cell culture
Heat inactivated	Avoids possibility of complement-mediated lysis – useful when culturing cells from the immune system
Iron supplemented	Improved growth rate in some cases
Lyophilized	Stable for long periods, reduced shipping costs
Irradiated	Extra guarantee of sterility
Dialyzed 1000 mol. wt cutoff	Selective removal (reduction) of components based on pore size; useful for radiolabeling studies
Dialyzed 10 000 mol. wt cutoff	As above – but additional removal of higher molecular weight products
γ-globulin depleted	Monoclonal antibody production
Charcoal stripped	Alternative product removal system, includes removal of lipid-soluble factors (steroid hormones, thyroid hormones) such as estrogen
De-lipidated	Lipid depleted, by fumed silica, reduced levels of cholesterol and triglycerides, may thus include removal of lipid-soluble contaminants, i.e. glucocorticoids, retinoic acid, etc.
Ultracentrifuged	Additional viral depletion
Reduced fibronectin	Extracellular matrix (ECM) related studies
β-propiolactone treated	Additional viral inactivation

Table 2. Some commercially available serum-free media

Cell type	Media base (supplier)	Additions (where stated)
Keratinocyte	MCDB 153 (C,G,S)	0.5 μg ml^{-1} hydrocortisone 5 μg ml^{-1} insulin 30 μg ml^{-1} bovine pituitary extract (BPE) 0.1 ng ml^{-1} EGF
Melanocyte	MCDB 153 (C)	0.5 μg ml^{-1} hydrocortisone 5 μg ml^{-1} insulin 26 μg ml^{-1} BPE 1 ng ml^{-1} fibroblast growth factor (FGF) 10 ng ml^{-1} phorbol ester
Chinese hamster ovary (CHO) cell	MCDB 302 (G)	Insulin Transferrin Methylcellulose
Macrophage	(G)	—
Endothelial cell	MCDB 131 (C,G,S)	1 μg ml^{-1} hydrocortisone 12 μg ml^{-1} bovine brain extract 10 ng ml^{-1} EGF
T lymphocyte	(G)	Insulin Transferrin Albumin

Continued

Supplements to Media

Table 2. Some commercially available serum-free media, *continued*

Cell type	Media base (supplier)	Additions (where stated)
Hybridoma	Ham's F12 (G,I,S)	Insulin Transferrin Cholesterol Bovine serum albumin Progesterone
Fibroblasts	MCDB 202 (C,S)	1 ng ml^{-1} FGF 5 µg ml^{-1} insulin
Bronchial/tracheal epithelial cells	LHC-9 (C)	0.5 µg ml^{-1} hydrocortisone 0.5 ng ml^{-1} EGF 52 µg ml^{-1} BPE 0.5 µg ml^{-1} epinephrine 10 µg ml^{-1} transferrin 5 µg ml^{-1} insulin 0.1 ng ml^{-1} retinoic acid 6.5 ng ml^{-1} triiodothyronine
Mammary epithelial cells	MCDB 170 (C)	0.5 µg ml^{-1} hydrocortisone 52 µg m^{-1} BPE 10 ng ml^{-1} EGF 5 µg ml^{-1} insulin
Skeletal muscle cells	MCDB 120 (C)	10 ng ml^{-1} EGF 100 µg ml^{-1} insulin 500 µg ml^{-1} fetuin 0.39 µg ml^{-1} dexamethasone

Media suppliers: C, Clonetics; G, Gibco-BRL; I, ICN-Flow; S, Sigma.

Table 3. Antibiotics

Antibiotic	Recommended concentration	Spectrum	Storage (°C)	Stability (37°C) (days)
Amphotericin B	2.5 mg l^{-1}	Fungi/yeast	4	3
Ampicillin	100 mg l^{-1}	+/− Bacteria	4	3
Chloramphenicol	5 mg l^{-1}	Bacteria	20	5
Chlortetracycline	5 mg l^{-1}	+/− Bacteria	0	1
Ciprofloxacin	10–40 mg l^{-1}	Mycoplasma	20	—
Dihydrostreptomycin	100 mg l^{-1}	+/− Bacteria	4	5
Gentamicin	50 mg l^{-1}	+/− Bacteria, mycoplasma	4	5
Kanamycin	100 mg l^{-1}	+/− Bacteria, mycoplasma	4	5
Neomycin	50 mg l^{-1}	+/− Bacteria	4	5
Nystatin	50 mg l^{-1}	Fungi/yeast	0	3
Penicillin	100 U ml^{-1}	+ Bacteria	20	3
Polymyxin B	50 mg l^{-1} 100 U ml^{-1}	− Bacteria	4	5
Spectinomycin	20 mg l^{-1}	+/− Bacteria	4	—
Streptomycin	100 mg l^{-1}	+/− Bacteria	4	3
Tetracycline	10 mg l^{-1}	+/− Bacteria	0	4
Tylosin	8 mg l^{-1}	+/− Bacteria, mycoplasma	4	3

+/−, Gram-positive/Gram-negative.
These are suggested concentrations only, and some cell types may be more sensitive to particular agents than is generally the case.

Table 4. Growth factors/cytokines/hormones

Factor	Concentration	Applications and references
BPE[a]	25–100 µg ml^{-1}	Aqueous extract from bovine pituitaries containing, among others, basic FGF, hormones and attachment factors; originally developed as a mitogen for vascular endothelial cells. Stimulates the growth of human prokeratinocytes and thymic, bronchial and mammary epithelial cells. Also used as defined supplement in some serum-free media formulations [3]
ECGS	50–200 µg ml^{-1}	Tissue extract from either bovine neural tissue, pituitary glands, whole brain or hypothalamus. Mitogenic for vascular endothelial cells. Used as a hybridoma growth factor eliminating the feeder cell requirement. Used widely as a supplement with and without serum for specific cell types, including human keratinocytes. Opinions divided on whether it contains FGF and ECGF since others say mitogenic activity physically, chemically and biologically distinct from FGF effects, see also BPE [4–6]
ECGF	2–10 ng ml^{-1} + 5 U ml^{-1} heparin	Human endothelial cell growth factor is homologous to acidic FGF with an additional N-terminal extension of 15 amino acids. Originally isolated from bovine brain as a mitogen for human endothelial cells. Mitogenic for mesodermally and neuroectodermally derived cells, including fibroblasts and vascular endothelial cells. The α-ECGF results from the cleavage of 20 amino acids from the N-terminus of the β-ECGF. As with type I heparin binding growth factors, its activity is greatly potentiated by heparin. ECGF is available as the recombinant human protein expressed in *E. coli*. ECGF from bovine brain can be purchased [7, 8]
EGF	1–25 ng ml^{-1}	Mitogen for fibroblasts, epithelial, glial, keratinocyte, endothelial and chondrocyte cells, i.e. mesodermal and ectodermal cell types. Used in reduced serum and

serum-free culture conditions, and promotes colony formation in epithelial cells. Human and mouse EGFs are 70% amino acid homologous and highly cross-reactive. Male mouse submaxillary gland, human urine-extracted and recombinant (expressed in *E. coli* and yeast) EGFs are all readily available, along with some biologically active peptide fragments [9,10]

EPO	0.1–2 U ml^{-1}	Stimulates the differentiation and proliferation of erythroid progenitor cells to more mature red blood cells. Allows the *in vitro* culture of some factor-dependent human erythroleukemia cells. Encourages erythroid colony formation in semisolid growth media assays. Human erythropoietin available either from human urine, or as the recombinant protein expressed in *E. coli* or CHO cells with identical glycosylation to the natural product. Murine recombinant protein has also been expressed in CHO cells, again identical to the natural protein [11, 12]
FGF		A family of polypeptide growth factors, at the moment comprising seven members with 30–50% sequence identity, most commonly available as follows [13]
FGF acidic	1–25 ng ml^{-1} + heparin	Sequentially identical though shorter than ECGF, with a similar spectrum of target cell types. They stimulate the proliferation of all cells of mesodermal origin and many of neuroectoderm, ectoderm and endodermal origin. Thus, responsive cells include fibroblasts, keratinocytes, endothelial cells, astrocytes, oligodendrocytes, epithelial lens, smooth muscle cells and melanocytes. Optimum activity when used in conjunction with heparin. Available from numerous sources; choice includes bovine brain and pituitary, and human recombinant expressed in *E. coli*. In one case human recombinant supplied with porcine heparin for optimium activity; also some biologically active synthetic polypeptide fragments available. See also KGF [14]

Continued

Supplements to Media

Table 4. Growth factors/cytokines/hormones, *continued*

Factor	Concentration	Applications and references
FGF basic	0.5–20 ng ml^{-1}	55% homology to FGF acidic. Mitogenic *in vitro* for cell types, including mesenchymal, neuroectodermal and endothelial cells; many targets same as FGF acidic, and indeed use the same high-affinity receptor. Can replace or enable the reduced use of serum. Human source active in murine cell lines. Availability includes bovine pituitary, a 1–24 amino acid fragment from bovine brain, bovine recombinant (both 146 and 156 amino acids), and, finally, human recombinant [15]. See also KGF
G-CSF	1–5 ng ml^{-1} or 50–200 CFU ml^{-1}	Affects the proliferation, differentiation and/or activation of hematopoietic cells of the neutrophil/granulocyte lineage. Human G-CSF is 73% identical at amino acid level to that of the mouse and is also active on mouse cells; also has been shown to have significant homology with IL-6. Suggested concentrations differ according to whether growth or colony formation is to be assayed. Human recombinant protein is available, expressed in either *E. coli* or yeast [16–18]
GM-CSF	0.1–50 ng ml^{-1}	A growth and maturation of function factor for hematopoietic cells. Human and mouse species do not cross-react despite 54% amino acid homology. Initially characterized for its ability to promote neutrophil, monocyte/macrophage and eosinophil colony formation, although it has also been shown to affect some human endothelial cells, and some tumor cell lines. Activity measured by half maximal stimulation of proliferation of cell lines, or by colony formation; lower end of concentration range for growth, upper end for colony formation. Mouse and human recombinant proteins expressed in both *E. coli* or yeast; also both mouse and human activated T-lymphocyte derived factors available [19, 20]

HGF	10–50 ng ml^{-1}	The most potent mitogen known for hepatocytes, with morphogenic, mitogenic and growth inhibitory activity on endothelial and epithelial cell types. Species cross-reactivity is illustrated by its effect on canine (MDCK) cells and bovine pulmonary artery endothelial cells. Receptor has been demonstrated to be the oncogene c-*met*. Available from human placenta, or recombinant protein expressed in insect sf21 cells [21, 22]
IFN-α	0.1–10 ng ml^{-1}	Inhibitor of viral replication and proliferation of some tumor lines. As an immunomodulator can activate natural killer (NK) cells; suppresses lymphocyte proliferation but enhances immunoglobulin secretion. This IFN is released by monocytes and macrophages after activation by viruses or dsRNA. Stimulates differentiation in some cells while suppressing the cloning efficiency of myeloid stem cells. Human and mouse products do not cross-react. Human recombinant expressed in *E. coli* available, as well as purified human and mouse IFN from cell culture [23, 24, 26]
IFN-β	10–1000 U ml^{-1}	Enhances NK activity, and induces anti-viral activity, inhibits the growth of fibroblasts. Secreted after Poly IC treatment of many cell types including fibroblasts, human foreskin, epithelial, myoblasts and lymphoblasts. Available as the human recombinant protein from *E. coli*, and purified from the secretions of mouse or human virus-treated fibroblasts, and (human) osteosarcoma cells [25]
IFN-γ	0.1–10 ng ml^{-1} or 10–400 U ml^{-1}	Acts as an immunomodulator, inducing enhanced proliferation of NK cells and T lymphocytes. Induces Fc receptors on monocytes and macrophages, IL-2 receptors on T cells, major histocompatibility complex (MHC) class I and II expression, and promotes functional activity of macrophages and NK cells. Not as potent an anti-viral

Continued

Supplements to Media

Table 4. Growth factors/cytokines/hormones, *continued*

Factor	Concentration	Applications and references
		agent as IFN-α or -β. Synergistic anti-tumor activity with tumor necrosis factor (TNF), secreted by activated T cells and NK cells. Human and mouse recombinant protein expressed in *E. coli* available [26–28]
IGF-I	1–50 ng ml^{-1}	Potent mitogen, structurally related to insulin, can replace serum for some cell lines (MDBK, MCF-7). Receptor shows similarity with insulin receptor. Stimulates a range of mesenchymal cells, including primary rat chondrocytes, Balb/c 3T3 fibroblasts, hybridomas and CHO cells. Available in recombinant human form expressed in either *E. coli* or yeast, and also recombinant peptide fragments [29–31]
IGF-II	5–100 ng ml^{-1}	Insulin- and IGF-I-like activity; 76% sequence identity with IGF-I, cross-reacts with insulin and IGF-I receptors. *In vivo*, mainly responsible for fetal growth. *In vitro* mitogenic targets include breast tumor cells, glioma, hepatocytes, pituitary, chondrocytes, adrenal cells and fibroblasts. Can replace serum in some cell types. Availability: human recombinant expressed in either yeast or *E. coli*, and recombinant peptide fragments [29, 32]
IL-1 ($\alpha + \beta$)	50–400 pg ml^{-1} or 2–50 U ml^{-1}	Activates T lymphocytes to release lymphokines. B-lymphocyte maturation and proliferation. Also has effects on hepatocytes and synovial cells. The α and β forms share 62% amino acid homology, their effects are mediated through the same receptors, and they share much the same biological activity. Both human α and β forms are active in mouse cells but with reduced intensity. Available as human purifed α, β or a mixture. Both human and mouse α and β forms are available as recombinant proteins expressed in *E. coli*. Additionally, a human polypeptide β fragment is available that demonstrates some but not all of the natural biological

		action. Also available is a human recombinant receptor antagonist that can compete for binding to the type 1 receptor [33, 34]
IL-2	5–100 U ml^{-1}	Stimulates growth and lymphokine production of activated T cells. Induces cytotoxic T-lymphocyte (CTL) and lymphocyte-activated killer (LAK) cell activity. Produced by activated T cells, also has an effect on B-cell growth and differentiation. The wide range of suggested concentrations is a result of a wide range of effects, i.e. use upper end of range for LAK cells. Human IL-2 has 60% amino acid homology with mouse IL-2, and is active on murine cells. Human, mouse and rat natural IL-2 are available; in addition recombinant IL-2 expressed in *E. coli* and yeast (human and mouse), baby hamster kidney (BHK) cells (human) and CHO cells (rat). Additional choices include the biotinylated human recombinant protein, and minor amino acid changes to increase solubility (human) [35–37]
IL-3	Human, 1–10 ng ml^{-1}; mouse, 1–1000 U ml^{-1}	Primarily T-cell-derived factor. Promotes growth and/or differentiation of hemato-poietic stem cells and also more mature myeloid cells. The primary effect on early multilineage progenitors results in loss of responsiveness to IL-3 and increased sensitivity to other factors, i.e. GM-CSF, G-CSF, M-CSF and erythropoietin. The wide range of murine activites reflects different potential uses: 1–200 U ml^{-1} in liquid culture or 50–1000 U ml^{-1} for colony formation. Although mouse and human IL-3 show some homology, they do not cross-react in their biological activity. Available as human and murine recombinant proteins expressed in *E. coli* as mouse natural IL-3 from the conditioned medium of murine myelomonocytic WEHI-3 cells, and mouse IL-3 expressed in yeast. Human natural IL-3 constitutively secreted by the bladder carcinoma cell line 5637 [38–40]

Continued

Supplements to Media

Table 4. Growth factors/cytokines/hormones, *continued*

Factor	Concentration	Applications and references
IL-4	1–40 ng ml^{-1}	Has a broad range of effects on B and T cells, hematopoietic cell lineages and NK cells. Affects proliferation and differentiation of activated B cells, T cells and thymocytes. Induces class II MHC antigens and CD23 marker on B cells. Supports growth of some factor-dependent cell lines, i.e. human TF-1, or murine FDCP-1. Human and murine IL-4 do not cross-react in activity. Available in *E. coli*-expressed human and murine form, and murine form expressed in yeast [41, 42]
IL-5	1–100 U ml^{-1} 0.5–1 ng ml^{-1}	Stimulates B cells, thymocytes and (especially human) eosinophils. Glycosylated disulfide-linked homodimer, with 70% amino acid similarity between human and mouse. Both species cross-reactive with each other, although with differing major target cell types. The wide range of concentration reflects a wide target range, from human bone marrow colony formation, to proliferation of factor-dependent human TF-1 cells (bottom of the range). Available in human and mouse recombinant forms expressed in insect cells, with additional choice of human recombinant form expressed in *E. coli* or yeast, and mouse form in COS cells (see Chapter 5) [43, 44]
IL-6	1–50 ng ml^{-1} 50–100 U ml^{-1}	Produced by a wide variety of cells, including activated T cells, monocytes, hepatocytes and some tumor cell lines. Acts on a wide variety of processes in many cell types, including plasmacytoma growth, secretion of IL-2 and cytotoxic differentiation of T lymphocytes, hybridoma cell growth in the absence of feeder cells, and the differentiation of B lymphocytes (maturation of B cells to antibody producers). Mouse and human IL-6 share only 42% amino acid homology, although the human form is active on mouse cells; opinion seems divided on the effect of

mouse IL-6 on human cells (consensus is that the mouse may be 5–10 times less effective on human cells). Availability includes human natural (lymphocyte-derived), human recombinant IL-6 expressed in *E. coli*, yeast or CHO cells, and mouse recombinant IL-6 expressed in *E. coli* or insect cells. Also available is a recombinant human soluble IL-6 receptor, thought to be involved in the regulation of cytokine activity *in vivo* [45, 46]

IL-7	0.1–10 ng ml^{-1}	Originally isolated as a factor inducing proliferation of murine pre-B cells, later found to be a growth and differentiation factor for T cells *in vivo*. Acts as a cofactor in the stimulation of both T and B cells and synergizes with SCF, concanavalin (Con A) or phytohemagglutinin (PHA) in lymphocyte activation/differentiation. Has also been shown to induce differentiation of precursor cells to mature CTLs and LAK cells. Although the human and mouse forms share 60% amino acid homology, but the human form is active on mouse cells the converse is not thought to be the case. Availability is limited to human and murine recombinants expressed in *E. coli* [47,48]
IL-8	5–300 ng ml^{-1}	Produced by a wide variety of cell types, including monocyte/macrophages, fibroblasts, endothelial cells, keratinocytes, astrocytes, lymphocytes, etc., in response to inflammatory stimuli such as IL-1, TNF, LPS, virus, etc. Potent neutrophilic chemotactic and activating factor, also enhances neutrophil adhesion to endothelial cells, up-regulates complement receptor expression, chemotactic for basophils, T cells (in some conditions) and eosinophils, angiogenic both *in vivo* and *in vitro* and co-mitogenic for keratinocytes. Two main forms (termed monocyte-derived and endothelial-cell-derived) differ only in a 5 amino acid addition to the core 72 amino acids at the N-terminal end in the endothelial-cell-derived form; cleavage to the short 72 amino acid version increases activity *in vitro*. Founder member of the

Continued

Supplements to Media

Table 4. Growth factors/cytokines/hormones, *continued*

Factor	Concentration	Applications and references
		chemokine family (pro-inflammatory cytokines) characterized by four conserved cysteine (C) residues, that includes in the C–X–C subfamily IL-8, GRO/MGSA (GROα gene from hamster – was found to be identical to the human MGSA (melanoma growth-stimulating activity) gene, platelet factor 4, β-thromboglobulin and IP-10, and in the β, or C–C, family RANTES, MIP-1α, MIP-1β, MCP-1, etc. IL-8 is available in the human recombinant-derived monocyte or endothelial-derived form expressed in *E. coli*. The murine form is not commercially available at present; however, human IL-8 has been shown to be active on a wide range of species, including the mouse [49–51]
IL-9	0.01–5 ng ml^{-1}	T-cell-derived growth factor that enhances survival of primary T-cell lines, and acts on mast cells, murine fetal thymocytes and on erythroid progenitors (FDCP-Mix) in synergy with erythropoietin, IL-3 or GM-CSF. May act in some cases by inducing differentiation/survival to factor (i.e. IL-3) responsiveness. Human and mouse IL-9 share 56% homology at the amino acid level, and whereas the murine form is active on human cells, the human form is not active on mouse cells. Availability includes recombinant mouse and human IL-9 expressed in insect cells, and the murine form expressed in *E. coli* [52, 53]
IL-10	0.20–20 ng ml^{-1} ± IL-2, -3 or -4	Inhibitory/stimulatory factor for cells of the immune system. Originally detected as a factor produced by a murine T-helper subset that inhibited cytokine secretion in other subsets. Secreted by some B cells, Epstein–Barr virus (EBV)-immortalized B-cell lines, LPS-activated human monocytes, keratinocytes (in mice), etc. Inhibitory effect on IFN and IL-2 secretion by T-cell subsets; this effect may be mediated via co-cultured antigen-presenting cells. Can in certain circumstances down-regulate MHC

class II expression and suppress TNF release from macrophages, also deactivating inflammatory macrophage responses. Stimulatory effects are seen in co-stimulation with IL-2 and IL-4 on the proliferation of mouse thymocytes, and with IL-3 and IL-4 on some mast cell lines. Homology of 73% at the amino acid level is seen between human and mouse IL-10; the human form is active in mouse cells but not conversely. Also striking similarity with EBV open reading frame (ORF) named BCRF1, sharing some activities and sometimes called viral IL-10. Recombinants expressed in *E. coli* available for both mouse and human; recombinant human IL-10 also expressed in insect cells [54, 55]

| IL-11 | 0.1–10 ng ml^{-1} | Detected originally as a factor in the conditioned medium from a primate bone marrow cell line, distinct from IL-6, that nonetheless allowed the proliferation of an IL-6 dependent cell line. Has so far only been found to be secreted by adherent cells of mesenchymal origin such as stromal fibroblasts, fetal lung fibroblasts and trophoblasts. Has multiple effects on cells of both hematopoietic and other types, including effects in synergy with IL-3 and IL-4 on the proliferation of early myeloid progenitors; the inhibition of adipogenesis on 3T3-L1 pre-adipocytes; and, either alone or with other cytokines, has a stimulatory effect on erythroid progenitor cells. IL-6 and IL-11 show many similar effects but do not share a receptor, although they may exert their effect via a common downstream event. Commercial availability is limited to the human recombinant protein expressed in either insect cells or *E. coli*, although human protein is active on mouse cells [56, 57] |

| IL-12 | 0.1–5 ng ml^{-1} | Originally identified in the conditioned medium from EBV-positive RPMI-8866 B cells. Comprised of two subunits, one of which is similar to IL-6 and G-CSF, and the other shows similarity to the soluble IL-6 receptor. The major effects of this cytokine are |

Continued

Supplements to Media

Table 4. Growth factors/cytokines/hormones, *continued*

Factor	Concentration	Applications and references
		on NK and T cells, promoting the growth of activated NK cells, and CD4+ and CD8+ T cells, and enhancing NK cytotoxicity, LAK-cell generation and proliferation of mitogen-activated human lymphoblasts. The human protein, although sharing up to 70% sequence identity with the murine protein, has not been shown to be active in the mouse system. Available as a human recombinant protein expressed in insect cells [58, 59]
IL-13	0.2–10 ng ml^{-1}	A very recent addition to the interleukin family, expressed in activated T lymphocytes. Its activities include the regulation of monocyte function and human (although not mouse) B-cell proliferation, differentiation and immunoglobulin secretion. Human and mouse IL-13 (formerly designated P600) share 58% amino acid homology, and appear to cross-react. The murine protein stimulates proliferation of human erythroleukemia cell line, TF-1. IL-13 appears to share signal transduction components with IL-4. Its limited availability includes both human and murine recombinants expressed in *E. coli* cells [60, 61]
Insulin	1–10 μg ml^{-1}	Produced in B cells of the islets of Langerhans, two polypeptide chains (A, 21 amino acids; B, 33 amino acids) linked together by two disulfide bridges. Many and various effects *in vivo*. *In vitro* acts as a growth factor for many cells in serum-free medium, also used in serum-containing medium for some primary cell cultures. Availabliity includes human, bovine, porcine, ovine and equine pancreatic insulin, and human recombinant insulin expressed in either yeast or *E. coli* [62]
KGF	10–100 ng ml^{-1}	Member of the fibroblast growth factor family. 39% and 37% sequence identity with basic and acidic FGF, respectively. Originally detected when secreted by embryonic

fibroblasts, synthesized primarily by mesenchymal cell lines (and *in vivo* the stromal fibroblasts of many epithelial tissues). Acts as a potent mitogen for keratinocytes and epithelial cells, but has little or no activity on fibroblasts and endothelial cells. Native human protein (163 amino acids, heparin binding), acts on human and mouse cells, although the recombinant human protein can be up to 10 times more effective than the native on some mouse cells (BALB/MK). Limited availability appears to be restricted to the human recombinant protein expressed in *E. coli* [63, 64]

LIF 1–5000 U ml⁻¹

A multifunctional cytokine purified from the conditioned media of L929 cells; induces differentiation/reduced clonogenicity in murine M1 cells. Expressed by a variety of cell types, including activated T lymphocytes, monocytes, brain glial cells, thymic epithelia, liver fibroblasts and bone marrow stroma, and can be either matrix bound (minority) or diffusible. Activities overlap with IL-6, IL-11, and OSM and include inhibition of leukemia cell growth (by induction of differentiation), synergism with IL-3 in the production of hematopoietic progenitor cells, stimulation of myeloblast proliferation and the suppression of pluripotent embryo stem cell differentiation, etc. The wide range in suggested concentration reflects the wide target range, towards the bottom of the scale for induction of differentiation, and the top end for growth of embryo stem (ES) cells. Human and mouse forms are highly homologous, with nucleotide identity approaching 90% in places, and whereas the human is active on mouse cells, the murine is approximately 1000-fold less active in human cells. Present availability includes human and murine recombinant LIF expressed in *E. coli* [65–68]

Liver cell growth factor 10–100 ng ml⁻¹

First isolated from human plasma, disparate effects include growth stimulation in hepatoma, thyroid follicular cells, and lymphocytes; acts a viability/survival factor for eosinophils, liver cells, organ cultures and fresh tumor cells *in vitro*, a

Continued

Supplements to Media

Table 4. Growth factors/cytokines/hormones, *continued*

Factor	Concentration	Applications and references
		chemotactic factor for macrophages and mast cells, and can inhibit growth of L929 fibroblast cells. Also thought *in vivo* to be important in angiogenesis and wound healing, stimulating collagen synthesis in a number of fibroblast cell lines. The origin of the naturally occurring tripeptide is still something of a mystery, although some, if not all, of its effects may be mediated by its ability to facilitate the uptake of Cu^{2+} from the surrounding medium. Available as a synthetic peptide [69–71]
M-CSF	50–200 Cfu ml^{-1}	Stimulates the formation of macrophage colonies in bone marrow hematopoietic progenitor cells. Prolongs *in vitro* survival and enhances activity of mature neutrophils, has chemotactic activity for granulocytes, monocytes and some mesenchymally derived cells. The 73% homology between mouse and human is reflected in the cross-reaction of biological activity. Human recombinant M-CSF expressed in either *E. coli* or yeast is available [72, 73]
NGF	0.1–20 ng ml^{-1} (β and 2.5S) or 1–100 ng ml^{-1} (7S)	A complex protein composed of five subunits in three classes (α, β and γ). 7S NGF is the entire complex and contains two α, one β and two γ subunits. The intact β subunit can be separated from the pentamer, consists of a homodimer held together by non-covalent bonds and is thought responsible for most NGF biological activity (known as 2.5S NGF). Finally, the single subunits comprising the 2.5S NGF can be purchased, these are referred to as β-NGF. There is some confusion in nomenclature here as some companies regard 2.5S NGF and β-NGF as the same thing. The 7S NGF has a molecular weight of around 140 000, whereas most regard the molecular weight of 2.5S NGF to be 26 000 and β-NGF to be 13 500–14 000 (i.e. roughly half of 26 000). β-NGF is fully active and the human and mouse proteins are

90% homologous and cross-reactive. NGF is secreted by (mouse) submandibular duct epithelium, neurons, fibroblasts, smooth muscle cells, astrocytes, etc., and affects a multitude of neuronal and other cell types, including survival of sensory neurons, and proliferation of adult adrenal chomaffin cells, outgrowth of chick dorsal root ganglia, chemotaxis for neutrophils, B-cell differentiation/proliferation. It synergizes with GM–CSF in the induction of basophil production and (NGF) degranulation, promotes growth of human erythroleukemic TF-1 cells and inhibits growth and promotes neurite outgrowth of murine pheochromocytoma (PC12) cells. Commercial availability: human recombinant β-NGF expressed in either *E. coli* or murine myeloma cells; murine 7S, 2.5S and β purified from mouse submaxillary gland; alternative 7S NGF also purified from bovine milk, extracted from *Vipera lebetina, Echis multisguamatus* and *Naja axiana* venom [74–76]

| OSM | 0.2–20 ng ml^{-1} | Secreted by phorbol myristate acetate (PMA)-treated human U937 monocytoid histiocytoma cells, activates T lymphocytes and monocytes. Inhibitory for human A375 melanoma cell proliferation, tumor lines such as HTB10 (neuroblastoma), MCF-7 (breast carcinoma) and endothelial cells, also upregulated low-density lipoprotein (LDL) receptors on hepatoma cells, and stimulates release of G-CSF, GM-CSF and IL-6 from human endothelial cells. Conversely, also found to stimulate the growth of AIDS-related Kaposi's sarcoma cells and fibroblasts. Shows similarity with LIF, G-CSF, IL-6 and CNTF (ciliary neurotrophic factor). Only available as a human recombinant expressed in *E. coli*, although human protein is active on murine cells [77, 78] |

| PDGF | 1–10 ng ml^{-1} | Said to be the principal mitogen affecting mesenchymal cells in serum. Originally found to be platelet-secreted, it has since been found to be secreted from many cell |

Continued

Supplements to Media

Table 4. Growth factors/cytokines/hormones, *continued*

Factor	Concentration	Applications and references
		types, including activated monocyte/macrophages, arterial endothelial cells and some tumor cell types. Used for culture of smooth muscle, fibroblasts, glial cells, neutrophils and primary human tumors. Also chemotactic for fibroblasts, and can stimulate neutrophil phagocytosis and collagen synthesis (among other things). Composed of two covalently linked (disulfide bridge) subunits (A and B, related by 60% protein homology), can be either AA or BB homodimers or AB heterodimer. Availability includes human natural PDGF (originally thought to be 70% A/B, may in fact be 70% A/A), porcine natural (B/B), and human recombinant A/A, A/B and B/B expressed in *E. coli*. Human and porcine PDGF are both active on mouse cells [79, 80]
SCF	1–100 ng ml^{-1}	Identified as the ligand for the *c-kit* oncogene, implicated in the regulation of hematopoiesis, melanogenesis and gametogenesis. Sources *in vivo* include bone marrow stromal cells, fibroblasts, liver cells, Sertoli cells and endothelial cells. Target cells mainly melanocytes, primordial germ cells, hematopoietic progenitor cells and mast cells. Can act synergistically (under some conditions) with any of the following: IL-3, IL-6, IL-11, G-CSF, in the induction of erythroid and myeloid colonies, or IL-7 in the clonal expansion of CD45+ B cells. Mouse and rat share approximately 80% amino acid identity with human; mouse and rat are active on human cells but the human factor is up to 800 times less effective on the rodent cells. The wide range of suggested concentrations reflects a wide range in usage. Available as recombinant human and murine proteins expressed either *E. coli* or yeast [81, 82]

TGF-α	1–10 ng ml^{-1}	Member of the EGF family (35–40% homology), and acts via the EGF receptor, although, unlike EGF, it may be more potent as an angiogenic and keratinocyte migration factor. Expressed by a variety of virally and chemically transformed cells, and also during embryogenesis. Synthesized by normal cells such as skin keratinocytes, macrophages, brain and pituitary tissue. Target cells include endodermal and ectodermally derived cells, and can reversibly confer a transformed phenotype (in conjunction with TGF-β) on nonneoplastic cells. Available as a human recombinant expressed in *E. coli*, human and rat synthetic peptides, and peptide fragments of human and rat forms [83, 84]
TGF-β	0.1–10 ng ml^{-1}	Five isoforms of TGF-β (1–5), the most commonly used being that originally described and derived from human platelets: TGF-β (now TGF-β1, also from myocytes, chondrocytes, astrocytes, epithelial cells and fibroblasts). A multifunctional homodimeric disulfide-linked factor, generally stimulating cells of mesenchymal origin, Schwann cells and osteoblasts and inhibiting lymphocytes, epithelial and endothelial cells. Closely related (70–80% sequence identity) to TGF-β2, -β3 -β4 and -β5 and more distantly (30–40%) to the activins, inhibins and BMP (bone morphogenesis proteins). Thought to be an important modulator of ECM secretion, *in vitro* can either inhibit (hepatocyte, lymphocyte, endothelial cell growth, IgG and IgM secretion from B cells, adipocyte and myocyte differentiation and monocyte respiratory burst) or activate (fibroblast and osteoblast growth, ECM secretion, helper T-cell function, monocyte cytokine production) cellular functions. *In vitro* β1, β2 and β3 appear roughly equivalent. Extremely well preserved across species (up to 90% sequence identity), resulting in good cross-reaction *in vitro*, synergistic in some cases with EGF.

Continued

Supplements to Media

Table 4. Growth factors/cytokines/hormones, *continued*

Factor	Concentration	Applications and references
		Availability:
		β – probably $\beta1 + \beta2 + \beta3$ – human platelet
		– pig platelet
		$\beta1$ – human and pig platelet
		– recombinant human in CHO cells
		– recombinant human in mammalian cells
		$\beta1.2$ – heterodimer
		$\beta1/2$ – pig platelet
		$\beta2$ – pig platelet, human recombinant in *E. coli*
		$\beta3$ – recombinant human in insect cells
		– recombinant chicken in insect cells
		$\beta4$ – not available
		$\beta5$ – recombinant *Xenopus* in insect cells [85, 86]
TNF-α	0.1–5 ng ml^{-1} or 100–5000 U ml^{-1}	Originally a factor secreted by activated macrophages, now also a product of activated T lymphocytes. TNF-α and -β are cytostatic/cytotoxic to a range of transformed cells, toxic to vascular endothelial cells, necrotic to tumor cells (murine Meth-A), can induce MHC class I and II expression, can activate polymorphonuclear leukocytes, have anti-viral activity and, with IFN-γ, induce differentiation in human neuro-blastoma cells. Although their activities have substantial overlap (they use the same receptors), TNF-α is normally thought to be the most active, and shares many of the characteristics of IL-1. Human and mouse TNF-α share 79% amino acid homology, and the human factor is active on mouse cells. Availability includes human natural

monocyte-derived, and recombinants expressed in yeast and *E. coli*; and mouse *E. coli* expressed recombinant; also a synthetic 17 amino acid peptide that retains at least some of the biological activity of the full-length protein. See also TNF-β below [87, 88]

TNF-β	0.05–50 ng ml^{-1}	Secreted by activated T cells, similar functions to the TNF-α, also synergistic with IFN-γ. Less potent than TNF-α, the human β protein shares 28% amino acid homology with the α (35% in mouse). TNF-β human and mouse proteins share 74% amino acid homology, and the human is active on mouse cells. Less widely available than the TNF-α, consisting of human recombinant expressed in *E. coli* [88]

[a]See *Table 5* for full names.

Recombinants expressed in *E. coli* are standard; however, the recombinants expressed in insect cells (usually *Spodoptera frugiperda* 21 cells, Sf21), BHK or CHO cells may be more confusing. Recombinants expressed in *E. coli* tend to differ from the natural proteins in the sugar side-chains that are added after translation. In some cases this does not matter; where it appears to do so then these proteins are often expressed in the higher eukaryotes, such as yeast (*Saccharomyces cerevisiae*), CHO (Chinese hamster ovary), BHK (baby hamster kidney) or COS-1 (African green monkey) that can more closely replicate the natural protein product. Expression of proteins in Sf21 cells uses the baculovirus expression system, which allows for very high levels of protein production to be achieved.

Supplements to Media

Table 5. Glossary of factor names, shortened forms and alternative names

BPE	bovine pituitary extract
EPO	erythropoietin
ECGS	endothelial cell growth supplement
EGF	epidermal growth factor, a.k.a. β-urogastrone (hEGF)
aFGF	fibroblast growth factor (acidic), a.k.a. FGF-1 or HBGF-1
bFGF	fibroblast growth factor (basic), a.k.a. FGF-2 or HBF-2
Liver cell growth factor	a.k.a Gly–His–Lys, H–Gly–His–Lys–OH, or glycyl–L-histidyl–L-lysine
G-CSF	granulocyte colony-stimulating factor, a.k.a. pluripoietin
GM-CSF	granulocyte/macrophage colony stimulating factor, a.k.a. CSF-2, pluripoietin, neutrophil migration inhibitory factor
HGF	hepatocyte growth factor, a.k.a. hepatopoietin A or scatter factor
IGF-II	insulin-like growth factor II, a.k.a. somatomedin A or MSA (multiplication stimulating activity)
IGF-I	insulin-like growth factor I, a.k.a. somatomedin C
IL-1α	interleukin 1, a.k.a. lymphocyte activating factor (LAF), BCDF, T-cell replacing factor, endogenous pyrogen, catabolin, or hematopoeitin 1
IL-1β	interleukin-1β
IL-2	interleukin 2, a.k.a. T-cell growth factor, thymocyte mitogenesis factor, T-cell replacing factor, or killer-helper factor
IL-3	interleukin 3, a.k.a. multi-CSF, CSF-2α, CFU-3A, mast cell/leukocyte growth factor, or P-cell stimulating factor
IL-4	Interleukin 4, a.k.a. B-cell growth factor (BCGF-1), T-cell growth factor (TCGF-2), mast cell growth factor, B-cell differentiation factor (BCDF), or B-cell stimulatory factor (BSF-1)
IL-5	interleukin 5, a.k.a. BCGF-II, EDF (eosinophil differentiation factor), TRF (T-cell replacing factor), IgA-EF (IgA enhancing factor), or colony stimulating factor eosinophil (CSF-Eo)
IL-6	interleukin 6, a.k.a. IFNβ2, BSF-2 (B-cell stimulatory factor 2), BCDF-2 (B-cell differentiation factor 2), hybridoma growth factor, HSF, PCT-GF, or IL-HP1
IL-7	interleukin 7, a.k.a. LP-1 (lymphopoietin 1)

IL-8	interleukin 8, a.k.a. NAF (neutrophil-activating factor), NAP-1 (neutrophil-activating peptide 1), MDNAP (monocyte-derived neutrophil-activating peptide), MDNCF (monocyte-derived neutrophil chemotactic factor), or LYNAP (lymphocyte-derived neutrophil-activating peptide)
IL-9	interleukin 9, a.k.a. TCGF III/P40 (T-cell growth factor III/P40) or MEA (mast cell enhancing activity)
IL-10	interleukin 10, a.k.a. cytokine synthesis inhibitory factor
IL-11	interleukin 11, a.k.a AGIF (adipogenesis inhibitory factor)
IL-12	interleukin 12, a.k.a. CLMF (cytotoxic lymphocyte maturation factor) or NKSF (natural killer cell stimulatory factor)
IL-13	interleukin 13, a.k.a P600 (murine)
KGF	keratinocyte growth factor, a.k.a. FGF-7 (fibroblast growth factor-7)
LIF	leukemia inhibitory factor, a.k.a. DIA (differentiation-inhibiting activity), DIF (differentiation inducing factor), D factor (differentiation stimulating factor), DRF (differentiation retarding factor), hepatocyte-stimulating factor III (HSF-III), CNF/CNDF (cholinergic neuronal differentiation factor), HILDA (human interleukin for DA cells), OAF (osteoclast-activating factor), or melanoma-derived lipoprotein lipase inhibitor (MLPLI)
M-CSF	macrophage colony-stimulating factor, a.k.a. CSF-1
NGF	nerve growth factor
OSM	oncostatin M
PDGF	platelet-derived growth factor, a.k.a. c-*sis*
SCF	stem cell factor, a.k.a KL (c-*kit* ligand), MGF (mast cell growth factor), or SLF (steel cell factor)
TGF-α	transforming growth factor α, a.k.a. sarcoma growth factor
TGF-β	transforming growth factor β, a.k.a. TIF-1 (tumor inducing factor-1) or CIF-A (cartilage-inducing factor) for β1 or CIF-B for β2 and probably DSF (decidual suppressor factor)
TNF-α	tumor necrosis factor α, a.k.a. cachectin
TNF-β	tumor necrosis factor β, a.k.a. lymphotoxin (LT)
IFN-α	interferon α, a.k.a. leukocyte interferon or type I interferon
IFN-β	interferon β, a.k.a. fibroblast interferon or type I interferon
IFN-γ	interferon γ, a.k.a. immune interferon or type II interferon

Chapter 4   **CULTURE VESSELS**

Cell culture requires a container that can be sterilized, does not leach toxic factors and also does not distort microscopy images; both plastic and glass fulfill these criteria. Adherent cells also require a surface to adhere to (also known as a substratum). Many companies supply the basic cell culture equipment, and for a number of years now the vessel type of choice has been a single-use disposable vessel, generally made from polystyrene, treated to encourage cell attachment [1], available in a variety of sizes and γ-irradiated for sterility. These 'standard' culture vessels are now available with growth areas ranging from 0.0133 cm^2 to 30 000 cm^2; choosing the ideal vessel can thus be complicated. Recent innovations, such as alternative types of plasticware (IVSP, PETG), expanded surface design (Corning Excell^{TM}, IVSP expanded surface), extracellular matrix (ECM) treated cultureware and the addition of 0.22 μm filters to the caps

surface area is available for adherence if the surface is corrugated rather than flat. If the corrugations are packed tightly enough (and there are enough of them) the available surface area can be doubled. It is important to remember, however, that the medium will become exhausted more quickly as more cells are seeded and cultured.

1.3 ECM treated
Anything from coverslips to 150 mm culture dishes can now be purchased (Beckton-Dickinson) precoated with a number of different attachment factors. These include collagen (rat and mouse), EHS ECM (Engelbreth–Holm–Swarm tumor extracellular matrix), laminin and poly-D-lysine, all of which can also be purchased individually (see Chapter 5, *Table 12*).

of flasks (i.e. Becton-Dickinson and Corning) mean that even the choice of design of culture vessel may now take some time.

1 Recent innovations

1.1 PETG

In Vitro Scientific Products (IVSP) (available through Imperial Laboratories Limited) market a plastic they term PETG, this is flexible and does not crack when dropped (they don't – I have tried it) and the vessels are molded in one piece. The products are claimed to give better cell growth characteristics, and the ability to cut away the top of the flask with a scalpel can be very useful.

1.2 Expanded surface (XPS) design

An idea that has recently caught on with a number of manufacturers, is based upon the fact that an increased

1.4 Filter caps

Tissue-culture flasks can be used in one of two ways:

1. In CO_2 incubators, a loose cap is used (usually using sodium bicarbonate buffering), to enable the media and CO_2 to be in contact. The water in the incubator base helps to keep the humidity high, so the medium does not evaporate. The disadvantage of this is the increased risk of contamination.
2. In dry incubators, the cap is tightened, usually using a Hepes-buffered medium. This is a less popular strategy, the disadvantage is that only airtight culture vessels can be used, limiting the type of culture that can be attempted. The positive side of this method is the reduced risk of contamination.

The use of the filter caps combines the best of both techniques. The pore size of 0.22 μm allows gas exchange but prevents contamination from outside.

The broad range of commercially available culture vessels used in most day-to-day culture is listed in *Table 1*. Seeding advice is based on calculations for 1×10^5 adherent cells in a 90 mm dish.

The list in *Table 1* represents guidelines on the use of the more common vessels. Of course there are many other, more specialized types of vessel to choose from, and these have not all been addressed here. For instance, the recommendation of culture volume in the 90 mm dish is based upon ease of use; however, up to 25 ml can be added to these plates, whereas theoretically 73.5 ml can be contained by the plate with 15 mm walls (in practice this would never be done). In addition, a range of depths of a 90 mm diameter dish can be purchased, from 10 mm right up to 25 mm.

Although not usually a problem with dishes, the additional volume in some of the extra-large flasks can cause problems with gas exchange. The maximum distance from the neck to the farthest end can increase quite substantially in the larger vessels while the surface area to volume ratio (crucial in gas exchange) can also drop dramatically.

Remember (when using XPS design) that the medium may be more subject to exhaustion of nutrients, medium color change is often a good guide to its condition (i.e. yellow – probably needs changing) (see Chapter 6, *Table 8*).

Each type of microbead/microsphere has its own directions for use, and additional information on the use of each is available from the supplier. Microcarrier beads are most commonly used in industry, where the beads have been under intensive investigation for their potential use in extracting large amounts of cell by-products from the spent medium. Much recent literature tends to reflect this [3,4]. However, general points to note for those in the lab include:

1. Loading microspheres is not always easy – cells will tend to adhere to the vessel rather than to the microsphere – cultures can be initiated from near-confluent roller bottles or by co-culture for a few days in stationary culture. A combination (partially stationary), stirring or rotating every half hour or so, can also be used.
2. High seeding densities will mean rapid medium exhaustion – some prefer fructose rather than glucose in the medium.
3. Be careful not to agitate the medium too violently, remember that these are not suspension cells, but

1.5 Microcarrier beads

Microcarrier beads (or microspheres) represent one of the more complex and sophisticated additions to the cell culture catalog [2]. Potentially these represent a relatively simple way to scale up the culture of adherent cells at low cost. These are now available from many companies and there are many variations.

The use of microbeads increases the surface area available for adherent cells to attach to, with little corresponding change in the volume of medium required (relative to the size of the culture vessel), as long as the beads remain suspended in the medium rather than being allowed to settle to the bottom of the flask. Spinner flasks and roller bottles, and beads designed to have the same density as the medium (i.e. neutral buoyancy), all keep the beads from settling to the bottom and causing localized confluence.

Beads are made of various materials, are different sizes, and may be treated in many different ways to encourage attachment. *Table 2* gives some examples of the types of beads now available.

adherent cells that just happen to be on a substratum that is in suspension.

4. Although harvesting the supernatant from these cultures is relatively easy (switch off the stirrer and allow to settle), harvesting the cells can be more problematic. Cells on microcarrier beads tend to be relatively resistant to release by trypsin (remember to wash the beads/cells free from medium before trypsinization, since serum will inhibit the trypsin). Sometimes the cells can be dislodged by agitation, or EDTA, but the best method is digestion of the bead itself. Thus, collagen beads are digested by collagenase, dextran beads are digested by dextranase and gelatin beads are digested by trypsin/EDTA.

5. As a general rule, the increase in cell number which you can expect will depend upon the increased surface area available, this in turn will depend upon the manufacturer's recommendations for the density of the beads per ml. The growth of cells on microspheres is in many ways similar to the growth of suspension cells, especially in that the cells can be sampled and the cell number determined without interfering with the whole culture.

Table 1. Small to intermediate size tissue-culture vessels

	Growth area (cm²)	Medium volume (ml)	Cell number × 10⁵ Adherent	Cell number × 10⁵ Suspension
Terasaki plate				
60–72 well	0.008	0.008	0.0016	0.016
Microwell plate				
96 well	0.32	0.3	0.0065	0.2
24 well	2	1	0.04	1–2
12 well	3.8	2	0.08	2–4
6 well	9.6	3.4	0.2	3–6
Culture dishes				
30 mm	6.9	2	0.14	4–8
35 mm	8	3	0.16	6–12
50 mm	17.5	4	0.36	8–16
60 mm	21	5	0.43	10–20
90 mm	49	10	1	20–40
100 mm	55	10	1.12	20–40
150 mm	145	20–30	2.95	60–100
Flasks				
25 cm²	25	5–10	0.5	10–20
75 cm²	75	15–30	1.5	30–60
175 cm²	175	50–100	3.5	100–200
225 cm²	225	up to 200	4.6	200–400

Table 2. Microcarrier/microsphere/microbead composition

Bead material	Outer coating	Density (g cm⁻³)	Diameter (μm)	Surface area (cm² g⁻¹)	Additional information
Dextran	DEAE	1.03	131–220	4400[d]	Primary cells, endothelial; charge throughout bead; require swelling (0.9% NaCl)
	DEAE	1.04	114–198	3300[d]	Mainly fibroblast cells; charge at surface only; require swelling (0.9% NaCl)
	DEAE	1.05	140–240	7000[d]	Require swelling in (0.9% NaCl)
	Collagen	1.04	133–215	2700[d]	Epithelial cells difficult to grow in culture; require swelling in 0.9% NaCl

Roller bottles				
1250 ml	490	100–200	10	200–400
2200 ml	850	100–250	17	200–500
4900 ml	1750	100–500	36	200–990
2300 ml (XPS)	1200	100–250	24.5	200–500

Plastic	Gelatin	1.02	95–150	—	γ-irradiated
	Gelatin	1.02	150–210	—	γ-irradiated
	Gelatin	1.03	95–150	—	γ-irradiated
	Gelatin	1.03	150–210	—	γ-irradiated
	Rapidcell™	1.03	150–210	—	
	Polystyrene[a]	1.04	150–210	350	Charged throughout
	Polystyrene[a]	1.05	160–300	255	Charged throughout
	Glass	1.03	150–210	350	Reusable
Gelatin	Gelatin[a]	1.03	150–250	4300	
	Gelatin	1.04	115–235	3800	
Glass	Collagen	1.02	90–150	—	
	Collagen	1.03	150–210	—	
	Glass[a]	1.02	95–150	—	Reusable
	Glass	1.02	150–210	—	Reusable
	Glass	1.03	95–150	—	Reusable
	Glass	1.03	150–210	—	Reusable
	Glass	1.04	95–150	—	Reusable
	Glass	1.04	150–210	—	Reusable
Collagen	Collagen[a]	—	100–400	—	

[a]Not a coating as such, but since the whole bead is made from the same material it acts in essentially the same way.
The superscript 'd' refers to the fact that the value of surface area to weight is for when the beads are dehydrated.

Chapter 5 **CELLS**

1 Continuous cell lines

The following chapter gives details of some of the cell types that can be obtained from the ATCC (American Type Culture Collection), the ECACC (European Collection of Animal Cell Cultures) or commercially (ICN-Flow, etc.), or are in general circulation. Included are recommended media, a few words of description, and (if not the original) a pertinent reference. The cell type list has been divided up on the basis of the major species represented – monkey (*Table 1*), human (*Table 2*), hamster (*Table 3*), mouse (*Table 4*), rat (*Table 5*) and other species (*Table 6*) – and alphabetically by cell name.

A word about media – some of the recommendations found in the catalogs are not those most often used. Many cells will (human) cells have a finite life span before becoming senescent.

In addition to these essentially normal cells, there are available (ATCC: genetic variant and normal human skin fibroblasts) human skin fibroblast cells derived from both apparently normal and genetically disordered and diseased patients (*Table 8*). They have been subjected to only a limited characterization in order to keep the passage number to a minimum. This valuable resource is, by its very nature, in limited supply. The cells have a population doubling expectancy of 15–35, and as few as 100 vials may be in existence.

The fibroblasts in the Detroit family were all isolated in the same way as part of a program with the ATCC [1] to provide

grow quite happily in a number of different media types, it is generally recognized that the closer one moves to the culture of cells in serum-free media, the more fastidious a cell becomes and the more crucial the correct medium becomes. In general, a combination of the advice from the supplier and a quick review of the current literature will be adequate to allow decisions on media.

Safety – many cells harbor and secrete live virus. Some cells are known to do this and should be treated accordingly. Almost certainly some cells not known to secrete infectious agents have become infected in the course of their culture.

2 Primary cells

The vast majority of human cell lines (i.e. continuous cells) are of tumor origin. Those that are not have been immortalized in some way, for instance EBV-immortalized B-lymphocytes. Some 'essentially normal' cell types are available (*Table 7*), these are not of tumor cell origin, are not transformed in any detectable way, and consequently a bank of cells from aneuploid individuals. Cells are primary, nontransformed and thus finite. Culture medium is as detailed in *Table 7* under Detroit 551. Note that Detroit 562 has not been included as it turns out to be epithelial-like in morphology.

The ECACC holds a bank of primary EBV-immortalized B-lymphocytes as a random cell line panel whose HLA profile has been determined. These cells have been immortalized and are not in such limited supply; on the other hand, they do not represent normal cells. The panel is in the process of being expanded up to 200 cell lines with the prospect of 500 in the future. Also the ECACC performs an 'EBV transformation service', where blood samples or lymphocytes can be sent to their laboratories at Porton Down. Further information on costs and postage, etc. should be requested from the ECACC (see Chapter 7).

Many primary cell types are thus readily available (see also Clonetics, supplied in the UK by Tissue Culture Services, who can supply, among other things, normal human

Cells

endothelial cells and epidermal melanocytes and keratino-cytes). There will, however, always be a need to originate new cells from fresh tissue.

Possible sources of such primary tissue are numerous, many strategies have been used, which mostly depend on the disaggregation or dissociation of the tissue. The enzymes used for tissue disaggregation are listed in *Table 9*. These enzymes are often most effective when used in combination and different combinations are recommended for different target tissues. Examples of enzyme protocols for different tissues can be useful as starting points in tissue disaggregation (*Table 10*).

Those enzymes that do not require either Ca^{2+} or Mg^{2+} (i.e. trypsin) can be used in conjunction with chelating agents (*Table 11*) [2]. These chelating agents can be used alone or in combination for either subculturing cells or for tissue disaggregation.

All of the chemical agents listed in *Table 11* can be purchased

culture vessel with attachment factors (*Tables 12* and *13*). Attachment factors are required by some specialized cells to intermediate between the cell surface and the substrate on which the cell is being cultured. Additionally, attachment factors are required by many cell lines when being cultured in reduced-serum or serum-free medium. Serum provides many factors, not fully defined as yet, which assist in attachment, and therefore its absence requires substitution with an appropriate factor. *In vivo* these factors are secreted by the cell as the 'extracellular matrix' which will anchor the cells to their basement membrane. It is now thought that attachment factors are also involved in cell growth and differentiation, with a role in the cellular interaction with growth factors.

Collagen. This factor has widespread use in culture. Type 1 (rat tail) is used for most normal and transformed mammalian cells (e.g. fibroblasts, hepatocytes, epithelial, endothelial and muscle cells). Type 1 (bovine and calf skin) is commonly used for primary epithelial cultures in addition to other cell types mentioned above. Type IV (mouse sarcoma)

from many commercial suppliers; in addition there are also commercial nonenzymatic cell dissociation buffers (Gibco-BRL) and cell dissociation solutions (Sigma). These may be used for either tissue disaggregation or subculturing (though they are not recommended for strongly adherent cells (Gibco-BRL). This may be useful if the cells are required, for instance, for cell surface receptor studies. These buffers/solutions are basically chelating agents made up in Ca^{2+}/Mg^{2+}-free balanced salt solution, with perhaps some additional elements, such as glycerol (Sigma). In many cases much the same result can be achieved with a 5–10 min incubation in versene.

3 Attachment factors

Once a tissue has been dissociated, the resulting released cells must be placed in the appropriate medium for their continued health. One major source of problems with primary cell culture is their attachment to the substratum. To overcome this problem many workers precoat the

is used for all epithelial, endothelial, muscle and neuronal cells. Type II (chicken) has been used for chrondrocyte culture. Type V (human) has been found to affect cellular functions of smooth muscle cells and endothelial cells.

Fibronectin. This is used under reduced and serum-free culture conditions for promoting attachment of fibroblasts and other mesenchymally derived cell types, such as human umbilical endothelial cells, kidney epithelial cells and sarcomas. As fibronectin has a broad range of activity it is commonly added to serum-free media formulations.

Gelatin. Gelatin is used for a variety of cell types and, like polylysine, it is a nonspecific attachment factor and is available on pre-coated culture vessels from some commercial suppliers. Gelatin has been used for culture of muscle cells, and the culture of cells in order to inhibit the growth of fibroblast cells.

Laminin. This is another factor which serves in an attachment role in addition to being a growth factor.

Laminin mainly affects endodermal and ectodermal cells (e.g. epidermal, neuronal, hepatocytes and secretory epithelial lines). It also has effects on Schwann cells (growth, migration and adhesion), muscle and various tumor cells. Laminin can also assist in fibroblast overgrowth in the culture of epithelial cells.

Poly-D-lysine. This is a synthetic attachment molecule and is used in culture for a number of cell types, including chick embryo cells and human amnion. Poly-D-lysine is often used in preference to poly-L-lysine to avoid the possibility of its breakdown to the biologically active amino acid, L-lysine, which could affect culture conditions by altering the amino acid balance of the medium.

Vitronectin. This molecule acts as ligand for some members of the integrin family of cell surface receptors. Vitronectin thus adsorbed on to the substratum can facilitate the attachment of cells possessing the appropriate receptors (e.g. endothelial, melanoma and fibroblast cells).

4 Fibroblast contamination

It is almost certainly true that cell lines of some type can be derived from any human. Explants and cultures of tumor samples from patients are more often than not contaminated with fibroblasts. This is usually a big problem, and many strategies have been utilized in order to remove these unwanted cells (*Table 14*).

Table 1. Selected continuous cell lines of monkey origin

Cell line	Recommended medium	Comments and references
B95-8	RPMI + 10% FCS	Marmoset suspension lymphoblastoid cells, leak infectious EBV after stressing. Popular choice as a convenient, easy to grow producer of infectious virus for EBV immortalization [3]
BS-C-1	EMEM + 10% FCS + NEAA	African green monkey kidney. Epithelial-like, adherent cells that show characteristic morphological changes after infection with SV40 virus. Also popular for transfection studies, especially with SV40-based vectors (see Chapter 5) [4]
COS-1	DMEM + 10% FCS	Derivative of CV-1. Transformed by origin of replication defective SV40. Single integration of early region including wild-type T antigen allows episomal maintenance of SV40-based plasmid vectors (see Chapter 5) [5]
CV-1	EMEM + 10% FCS	African green monkey kidney. Adherent, fibroblastic cells. Suitable host for SV40 infection. Show characteristic morphological changes and are popular as transfectable cells, especially for SV40-based vectors (see Chapter 5) [6]
LLC-MK2	199 + 1% HS + 1.6 g l^{-1} NaHCO$_3$	Rhesus monkey kidney epithelial cells. Original cells (Original) grew slowly and were thus adapted to EMEM + 5% FCS (Derivative) [7]
MLA 144	RPMI + 10% FCS	Lymphoma derived from spontaneous gibbon lymphosarcoma. Constitutive release of gibbon IL-2, which is biologically active in both mouse and human systems. Also releases gibbon ape leukemia virus and should be handled accordingly [8, 9]
Vero	199 + 5% FCS or EMEM + 5% FCS + NEAA	African green monkey kidney. Fibroblast-like. Popular choice for viral infection assays and culture of some human protozoan parasites [10–12]

RPMI, RPMI 1640; FCS, fetal calf serum; EBV, Epstein–Barr virus; NEAA, nonessential amino acids; EMEM, Eagle's MEM; DMEM, Dulbecco's modification of Eagle's medium; HS, horse serum; 199, Medium 199.

Cells

Table 2. Continuous cell lines of human origin

Cell line	Recommended medium	Comments and references
BeWo	Ham's F-12K + 15% FCS	Choriocarcinoma. The first human endocrine cell to be maintained in continuous culture. Epithelial-like, placental origin and secretes a spectrum of placental hormones. Adherent [13]
CCRF-CEM	RPMI + 10% FCS	Children's Cancer Research Foundation. T-lymphoblastoid suspension cells from a 4-year-old female with acute lymphoblastic leukemia [14]
CCRF-SB	RPMI + 10% FCS	B-lymphoblastoid suspension cells from a male child with acute lymphoblastic leukemia [15, 16]
Chang liver	BMEE + 10% FCS	One of the first cell lines serially propagated from nonmalignant tissue. Extremely useful in the study of malignancy. Adherent epithelial morphology [17]
Daudi	RPMI + 10% FCS	Burkitt lymphoma, B-lymphoblastoid EBV-positive suspension cell. Extensively used in the study of oncogenesis and interferon research [18]
Detroit 562	EMEM + 10% FCS +NEAA + pyruvate + Lac. Hy.	Derived from metastatic cells from a carcinoma of the pharynx. Adherent epithelial-like cells [19]
EB-3	RPMI + 10% FCS	Burkitt lymphoma, B-lymphoblastoid EBV-positive suspension cell. Male [20]
HeLa	EMEM + 10% FCS + NEAA	Cervical carcinoma. Epithelioid, adherent, later rediagnosed as adenocarcinoma, widely used cell line [21, 22]

HeLa 229	EMEM + 10% FCS + NEAA	Derived from HeLa cells, resistant to polio virus [21, 22]
HeLa S3	Ham's F-12 + 10% FCS	Derived from HeLa cells; readily adaptable to growth in suspension; plate with 100% efficiency [23]
Hep-2	EMEM + 10% FCS	Epithelial-like larynx carcinoma cells; supports the growth of 10 arboviruses and measles virus [24]
HL-60	RPMI + 10% FCS	Myeloid suspension cells derived from female patient with acute promyelocytic leukemia. Cells differentiate in response to treatment with a number of agents. Cells have been used for cloning myeloid cytokines [25, 26]
IM-9	RPMI + 10% FCS	B-lymphoblast suspension cells isolated from bone marrow of female patient with multiple myeloma [27–29]
K-562	RPMI + 10% FCS	Isolated from pleural effusion of a female patient in blast crisis of chronic myelogenous leukemia. Subsequent studies show that, although Philadelphia chromosome positive, the cells demonstrate both myeloid and erythroid markers [30]
KB	EMEM + 10% FCS + NEAA	Oral epidermoid carcinoma cells, early adherent monolayer culture [31]
KG-1	Iscove's + 20% FCS	Male erythroleukemia patient, suspension myeloid cells, differentiate into macrophage-like cells after treatment with phorbol ester [32]
MCF7	EMEM + 10% FCS + NEAA	Pleural effusion from female breast cancer post-radiation and hormone therapy, epithelial-like, adherent, characteristics of mammary epithelium [33]

Continued

Table 2. Continuous cell lines of human origin, *continued*

Cell line	Recommended medium	Comments and references
MOLT-4	RPMI + 10% FCS	Male acute lymphoblastic leukemia (ALL) relapse patient. Suspension T cells, no EBV or Ig detectable, able to rosette with sheep red blood cells (SRBC), TdT activity [34]
Namalwa	RPMI + 10% FCS	Burkitt lymphoma. Suspension B cells. Single X chromosome, no Y [35]
Raji	RPMI + 10% FCS	Burkitt lymphoma suspension B cells, male, EBNA positive, c-*myc* gene translocation and mutations [36]
RPMI 1788	RPMI + 10% FCS	EBNA-positive hematopoietic cells, isolated from the peripheral blood of an apparently normal male [37]
RPMI 2650	EMEM + 10% FCS + NEAA	Epithelial-like adherent cells isolated from nasal septum malignancy [38]
RPMI 8226	RPMI + 10% FCS	Hematopoietic cells isolated from a patient with multiple myeloma [39]
T24	McCoy's 5A + 10% FCS	Bladder transitional epithelium carcinoma. Epithelial-like adherent cells, used as target for cytotoxicity assays. Oncogene c-Ha-*ras* (mutant) cloned from this cell line. Secretes IL-6 [40]
WISH	BMEE + 15% FCS	Epithelial-like cells derived from human amnion tissue. Used as assay for measles virus virulence, supports polio virus and adenovirus type 3 [41]

BMEE, basal medium Eagle's with Earle's; Lac. Hy., 0.1% lactalbumin hydrolysate; TdT, terminal deoxynucleotide transferase; Pyruvate, 1 mM sodium pyruvate; EBNA, EBV nuclear antigen.

Table 3. Continuous cell lines of hamster origin

Cell line	Recommended medium	Comments and references
BHK-21 (C-3)	EMEM + 10% FCS + NEAA	Male baby Syrian hamster kidney cells clone C-3, adherent [42, 43]
CHO-K1	Ham's F-12 + 10% FCS	Proline-dependent, adherent epithelial-like cells isolated from Chinese hamster ovaries, widely used for mutation studies [44]
Don	EMEM + 10% FCS	Pseudodiploid Chinese hamster lung fibroblast-like cell. May also be grown in McCoy's 5A + 10% FCS, or Puck's N-16 + 20% FCS [45]

Cells

Table 4. Continuous cell lines of mouse origin

Cell line	Recommended medium	Comments and references
3T3 (A31)	DMEM + 10% FCS	Balb/c fibroblast. Similar preparation, growth and properties to Swiss 3T3, but from the inbred mouse strain Balb/c. See also NIH 3T3 [46]
3T3 (Swiss)	DMEM + 10% FCS	Fibroblast cell originally prepared from disaggregated random bred Swiss mouse embryos. Adherent, highly contact inhibited, widely used for oncogene research. See also NIH 3T3 [47]
Am12	DMEM + 10% NCS	Full name GP + envAm12, Amphotrophic[a]. See also GP + E-86 and PA317. Based on NIH/3T3, *gag* and *pol* have been placed on a different plasmid from the *env* gene and transfected separately into NIH/3T3. Thus there is little chance of recombination to produce infective helper virus [48]
B16-BL6	RPMI + 10% FCS	Variant of B16-F10 selected for an increased tendency to form metastatic secondary tumors [49]
B16-F10	RPMI + 10% FCS	Melanoma, C57/BL6 mouse, adherent fibroblast-like cells, can secrete melanin, MHC class I and class II negative, highly malignant [50]
BW5147.G.1.4	DMEM + 10% FCS	Suspension T lymphoma, subclone of BW5147 from AKR/J mouse. MHC class II (H-2k) and Thy-1.1 antigen positive. Resistant to 6-thioguanine (10^{-4} M), die in HAT. Used as fusion partner for T-cell hybrids [51, 52]
C3H/10T1/2	EMEM + 10% FCS	Contact-inhibited embryo fibroblast from C3H mice. Nontumorigenic, do not

		spontaneously transform, no endogenous leukemia or sarcoma virus detectable. Differentiate into myoblasts, used to clone the *myo D* gene [53, 54]
Clone M-3	Hams F-10 + 15% HS +2.5% FCS	Adapted to culture from a melanoma in C × DBA F_1 hybrid mice. Forms melanotic tumors in mice [55]
Ehrlich's ascites	NCTC 135 + 10% FCS	Full name Ehrlich–Lettre ascites, strain E derived from explant of 7-day tumor of Ehrlich carcinoma. Epithelial-like cells reported to grow in both C57BL and Webster Swiss mice [56]
ES-D3	DMEM + 15% FCS + 100 µM Mercap	Pluripotent embryo stem-cell line isolated from the blastocyst of 129/Sv + C/ + P mouse. Spontaneously differentiate into embryonic structures, can be maintained as undifferentiated cells by growth on feeder layers or medium containing LIF-treated cells. Can be transfected and knockouts can be achieved by homologous recombination (gene targeting). Embryonic stem cells injected into blastocysts can colonize the germ line for transgenic mouse production [57–59]
FDCP-1	RPMI + 10% FCS	Factor-dependent continuous cell lines, Paterson Laboratories. Diploid myeloid cells isolated from bone marrow culture of DBA/2 mice. IL-3 dependent for growth, non-malignant in syngeneic mice, capable of differentiation into neutrophils and macrophage-like cells [60]
FDCP-Mix	RPMI + 10% FCS	Factor-dependent continuous cell lines, Paterson Laboratories. Much the same cells as FDCP-1. Cells infected with *src*-carrying retrovirus, multipotent with erythroid, megakaryocytic, neutrophilic, macrophage, osteoclast and mast cell differentiation capabilities. Lines raised in F_1 hybrid mice of strains C57/BL6 and DBA/2 [61]

Continued

Cells

Table 4. Continuous cell lines of mouse origin, *continued*

Cell line	Recommended medium	Comments and references
GP + E-86	DMEM + 10% NCS	Based on NIH/3T3 fibroblasts. Ecotrophic[b] retroviral packaging cells. *Gag* and *pol* genes are on a different plasmid to the retroviral *env* gene. Transfection of these two plasmids created packaging cells with little chance of the packaging signal from the disabled vector becoming associated with the helper virus replication. See also PA317 [62]
HSDM1C1	Ham's F-10 + 15% HS + 2.5% FCS	Subclone from cell line derived from Swiss albino mouse fibrosarcoma. Secretes prostaglandin E_2; fibroblastic [63]
L929	EMEM + 10% FCS + NEAA	Connective tissue, commonly known as L cells. Fibroblast-like, one of the first cloned strains from one of the first cell strains established in continuous culture. Male. Popular choice in transfection studies [64]
L1210	DMEM + 10% HS	Suspension cells from lymphocytic leukemia in DBA/2 mice. Chemical carcinogen induced, used for screening chemical agents for cytotoxicity [65]
LL/2(LLC1)	DMEM + 10% FCS	Also known as Lewis lung carcinoma, adherent cells established from C57 mouse bearing primary Lewis lung carcinoma. Highly metastatic [66]
MOPC-31C	L-15 + 20% FCS	Mouse, oil-induced plasmacytoma cell. Balb/c plasmacytoma suspension cell induced in animals by mineral oil injection. Secretes mg quantities of IgG_1 [67, 68]
MPC-11	DMEM + 20% IHS	Derived from Merwin plasma cell tumor in Balb/c mice, suspension myeloma cell line, secretes IgG_{2b} [69]

NB41A3	Ham's F-10 + 15% HS + 2.5% FCS	Subclone of adherent C-1300 mouse neuroblastoma cells. Resemble mature neurons and secrete neurotransmitter synthetic enzymes [70]
NCTC 2472	NCTC 135 + 10% HS	Connective tissue, high tumor producing, fibroblast-like, derived from NCTC 1742(VII). Produces tumors in 65% of C3H/HeN mice [71]
NCTC 2555	NCTC 135 + 10% HS	As above, except that this strain is low tumor producing. Under the same conditions as NCTC 2472 the NCTC 2555 produces no tumors [71]
Neuro-2a	EMEM + 10% FCS	Strain A albino mouse spontaneous neuroblastoma. Adherent, neuron and amoeboid-like. Produces large quantities of microtubular protein [72]
NIH/3T3	DMEM + 10% FCS	Similar to 3T3 (Swiss) in origin, result of five serial cloning steps in order to isolate the clone most suitable for oncogenic transformation studies [73]
P3/NSI/1-Ag4-1 (NS-1)	DMEM + 10% FCS	Abbreviation (NS-1), Balb/c mouse myeloma clone. Nonimmunoglobulin-secreting clone of P3X63Ag8. Azaguanine (10^{-4} M) resistant, does not grow in HAT. Used as hybridoma fusion partner [74]
P3X63Ag8	DMEM + 10% FCS	Balb/c myeloma derived from P3K which was itself derived and established in culture from a Balb/c plasmacytoma (MOPC-21). Resistant to azaguanine, dies in HAT medium, parent of P3X63Ag8.635 and P3/NSI/1-Ag4-1(NS-1) [75]
P3X63Ag8.635	RPMI + 20% FCS	Balb/c mouse myeloma subclone of P3X63Ag8. Has lost the ability to produce heavy- or light-chain immunoglobulin. Azaguanine resistant and HAT sensitive, widely used as hybridoma fusion partner [76]

Continued

Table 4. Continuous cell lines of mouse origin, *continued*

Cell line	Recommended medium	Comments and references
P388D1	DMEM + 10% HS	Suspension chemically induced DBA/2 lymphoid neoplasm derived, monocyte–macrophage cell line. Subline P338D1(IL-1) secretes IL-1 [77]
P815	DMEM + 10% FCS	DBA/2 mastocytoma, suspension cells synthesize lysozyme and phagocytose latex beads, no ADCC effector activity, used as CTL target cells [78]
PA317	DMEM + 10% FCS	Derived from NIH/3T3 TK-, and transfected with plasmid-borne (pPAM3) retroviral packaging constructs and HSV-TK. Allows the replication and secretion of infective, disabled, amphotrophic retroviral particles [79]
Sp2/0-Ag14	DMEM + 10% FCS	Balb/c-based nonsecreting myeloma. A reclone of Sp2/HL-Ag, which is itself derived from Sp2/HLGK, which is derived from the fusion hybrid P3X63Ag8 × Balb/c spleenocytes with anti-SRBC activity. Ag-14 is resistant to azaguanine 20 μg ml^{-1} and dies in HAT; does not secrete or synthesize immunoglobulin [80]
Y-1	Ham's F-10 + 15% HS + 2.5% FCS	Initiated from the adrenal cortex of a male LAF1 mouse. Epithelial cells, secrete steroids [81]
YAC-1	RPMI + 10% FCS	A/Sn mouse lymphoma induced after inoculation of new-born animals with Moloney leukemia virus. Suspension cell, popular choice as NK target [82]

NCTC, National Cancer Institute, Tissue Culture Section; Mercap, mercaptoethanol; L-15, Leibovitz's L-15 medium; IHS, heat-inactivated horse serum; NCS, new-born calf serum; HAT, hypoxanthine, aminopterin, thymidine; LIF, leukemia inhibitory factor; ADCC, antibody-dependent cell cytotoxicity; CTL, cytotoxic T-lymphocyte; SRBC, sheep red blood cells; HSV-TK, herpes simplex virus thymidine kinase.
[a]Amphotrophic, able to infect across the species barrier.
[b]Ecotrophic, only able to infect the same species.

Table 5. Continuous cell lines of rat origin

Cell line	Recommended medium	Comments and references
C_6	Ham's F-10 + 15% HS + 2.5% FCS	Chemically induced rat glial tumor cell. Fibroblastic. Secretes S-100 protein [83]
FRTL5	C.Ham's F-12 + 5%FCS	Fischer rat thyroid line 5, subline of FRTL, epithelial, adherent, requires the addition of extra supplements for growth *in vitro*: 10 μg ml^{-1} insulin, 10^{-8} M hydrocortisone, 5 μg ml^{-1} transferrin, 10 ng ml^{-1} Glycyl-L-histidyl-L-lysine acetate, somatostatin 10 μg ml^{-1}, thyroid stimulating hormone (TSH) 10 mU ml^{-1} [84, 85]
GH_3	Ham's F-10 + 15% HS + 2.5% FCS	Wistar rat pituitary tumor epithelial cell line. Secretes growth hormone and prolactin, both regulated by hydrocortisone. Similar to GH1 [86, 87]
Jensen sarcoma	McCoy's 5A + 5% FCS	Fibroblast-like rat sarcoma, requires asparagine, although independent variants occur at a rate of 1 : 10 000 [88]
LLC-WRC 256	199 + 5% HS	Highly malignant epithelial-like cells isolated from Walker rat carcinoma maintained in Harlan-Wistar rats. Supports propagation of herpes simplex and vaccinia virus [7]
MH1C1	Ham's F-10 + 15% HS + 2.5% FCS	Epithelial-like buffalo rat hepatoma. Adherent, synthesizes and secretes rat serum albumin [89]
NRK-49F	DMEM + 5% FCS + NEAA	Normal rat kidney fibroblast. Contact inhibited, sensitive to transformation by viral or chemical agents [90]

Continued

Cells

Table 5. Continuous cells of rat origin, *continued*

Cell line	Recommended medium	Comments and references
NRK-52E	DMEM + 5% FCS + NEAA	Normal rat kidney epithelial-like cells. Distinct, but from the same source as NRK-49F [90]
Y3-Ag 1.2.3	DMEM + 5% FCS	Lou rat myeloma, subclone of 210.RCY3.Ag1. Derived for the production of rat × rat hybridomas. Resistant to azaguanine (dies in HAT) [91]
YB2/0	DMEM + 10% FCS	Full name YB2/3HL.P2.G11.16Ag.20. Derived from YB2/3HL followed by selection for the absence of immunoglobulin secretion. YB2/3HL resulted from the fusion of Y3-Ag 1.2.3 with AO spleen cells [92]

C.Ham's F-12, Coon's modified Ham's F-12; 199, Medium 199.

Table 6. Continuous cell lines from other species

Cell line	Recommended medium	Comments and References
CRFK	EMEM + 10% HS + NEAA	Cat kidney epithelial cells, used extensively for cat viral studies [93]
LLC-PK1	199 + 3% FCS	Pig kidney epithelial cells, synthesize plasminogen activator [94]
LLC-RK1	199 + 10% HS + 1.12 g l^{-1} NaHCO$_3$	Rabbit kidney epithelial-like cells. Pooled New Zealand white animals, susceptible to viral infection (rubella, vaccinia, herpes) [95]
MDBK (NBL-1)	Ham's F-12 + 10% FCS	Bovine viral diarrhea free bovine kidney epithelial-like cells [96]
MDCK (NBL-2)	EMEM + 10% FCS	Maden Darby canine kidney epithelial cells. Used for veterinary vaccine production. Respond to hepatocyte growth factor by tubule formation in collagen gels [1, 97, 98]
MiCl1(S+L−)	RPMI + 10% FCS	Mink lung cell line selected after Moloney murine sarcoma virus (MSV) infection. Contains an MSV genome rescuable by superinfection with a compatible helper virus. Used as a sensitive bioassay for helper virus. Adherent, fibroblast-like [99]
Muntjac	Ham's F-10 + 20% FCS	Indian male muntjac deer skin cell line. Primarily of interest because of the low chromosome number – 7 male, 8 female [100]
SIRC	EMEM + 10% FCS + NEAA	Rabbit cornea, little information about first 400 passages. Fibroblastic, distinct cytopathic effects seen after infection with rubella [101]

Cells

Table 7. Finite cell lines of human origin

Cell type	Recommended medium	Comments and references
Detroit 551	EMEM + 10% FCS + NEAA + Pyruvate + Lac. Hy.	Skin fibroblasts from Caucasian female embryo. Finite life span of 25 serial subcultures. Normal karyology. Available at passage 8 [1]
IMR-90	EMEM + 10% FCS	Diploid embryonic lung fibroblasts [102]
MRC-5	BMEE + 10% FCS	Diploid, male, fetal lung fibroblasts. Can attain up to 46 population doublings. Susceptible to a wide range of human viruses. Obtainable at passage 10. Popular choice as a feeder layer for primary cultures [103]
MRC-9	EMEM + 10% FCS + NEAA	Diploid, female, fetal lung fibroblast-like. Average life span 44 doublings. Susceptible to a wide range of human viruses [104]
WI-38	BMEE + 10% FCS	Diploid, female, embryonic (3 month) lung. Unusually sensitive to pH (must be less than 7.4). Fibroblast-like, capable of up to 60 population doublings with 24 hour doubling time [105]

Pyruvate, 1mM sodium pyruvate; Lac. Hy., 0.1% lactalbumin hydrolysate.

Table 8. Detroit fibroblast cells

Cell name	Disorder/genetic lesion
Detroit 510	Female, galactosemia, apparently normal karyotype
Detroit 525	Female, Turner's syndrome, missing X
Detroit 529	Female, Down's syndrome, trisomy in a group G chromosome and in X
Detroit 532	Male, Down's syndrome, trisomy 21
Detroit 539	Female, Down's syndrome, trisomy group G
Detroit 548	Female, translocation, translocated D on partial extra D
Detroit 551	Female, normal diploid
Detroit 573	Female, translocation B/D

See text for further information. The culture medium which should be used is as shown in *Table 7* for Detroit 551.

Cells

Table 9. Enzymes for tissue disaggregation

Collagenase (EC 3.4.24.3)	From *Clostridium histolyticum*. Working conc. 0.05–0.25%, 0.075–3.75 U ml^{-1} in HBSS. Activator Ca^{2+}, inhibitors EDTA, EGTA, not inhibited by serum. A,B,D,H,P readily available. Consist of different crude preparations with different ratios of proteolytic activity, such as clostripain and neutral protease. A, balanced ratio; B, high collagenase, highish clostripain; D, normal to high collagenase, low trypsin; H and P are functionally assayed for activity in the preparation of hepatocytes (H, balanced enzyme ratio) and pancreatic islets (P, very high collagenase activity, and high clostripain activity) from the rat. Reconstitute stored at $-20°C$, avoid repeated freeze–thaw [106]
Collagenase (EC 3.4.24.8)	From *Vibrio alginolyticus*, much the same as above. Working conc. 0.1–0.25% (2–5 U ml^{-1}), often highly purified and used in conjunction with other enzymes such as dispase. Not inhibited by serum. Also from crab hepatopancreas – active against native insoluble collagen, and collagenase from *Hirudo medicinalis* (medicinal leech)
Dispase (EC 3.4.24.4)	From *Bacillus polymyxa*. Rapid and effective but also gentle nonspecific neutral protease. Activators include Ca^{2+}, Mg^{2+}. Inhibitors include EDTA, EGTA, Hg^{2+} and other heavy metals. Working conc. 0.5–3.0 U ml^{-1} (Ca^{2+}/Mg^{2+}-free PBS). Can also be used to prevent clumping in suspension cells, not inhibited by serum. Also known as protease in some catalogs. Beware: units of activity are often from different assay systems. Store reconstituted at $-20°C$ and avoid repeated freeze–thaw [107]
DNase I (EC 3.1.21.1)	From bovine pancreas. Activity > 2000 U mg^{-1} assayed on DNA. Double-stranded specific endonuclease. Reconstitute in PBS or balanced salt solution. Optimum activity in 5 mM Mg^{2+}. Working conc. 0.01–1 mg ml^{-1} (20–2000 U ml^{-1}). Inhibited by EDTA, EGTA and denatured by SDS. Cell lysis during tissue dissociation can release high molecular weight DNA which can cause cell clumping [108]
Elastase (EC 3.4.21.36)	From porcine pancreas. Working conc. 0.1–0.5 mg ml^{-1}, 10–50 U ml^{-1}. Reconstitute in physiological Tris

(pH 8.5) or sodium carbonate buffer (pH 8.8). Inhibited by PMSF, DFP, 1-antitrypsin, 2-macroglobulin. Degrades connective tissue fibre protein (elastin), also hemoglobin, fibrin, casein, albumin, soybean protein and denatured collagen. Preferential cleavage adjacent to neutral amino acids. Used to aid the release of cells from connective tissue, often in conjunction with trypsin and collagenase. Also from human neutrophil (EC 3.4.21.37) [109]

Hyaluronidase	Bovine testes (EC 3.2.1.35). Other sources include ovine testes and *Streptococcus hyalurolyticus*. Working conc. 0.1–1 mg ml^{-1} (100–1000 U ml^{-1}). Reconstituted in balanced salt solution. Activated by polycations, inhibited by serum, Ca^{2+}, Fe^{2+}, Fe^{3+}, heparin. Degrades the glycoaminoglycans hyaluronic acid and chondroitin. If used with protease, such as trypsin, can digest extracellular matrix [110]
Pancreatin	From porcine pancreas; natural mixture of pancreatic enzymes, including amylase, trypsin, lipase ribonuclease and trypsin. Can be 1–5 × the activity described by the N.F. and the USP. 1x will hydrolyze 25 × its weight in protein, 38 × its weight in starch, 114 × its weight in fat. Reconstitute in Ca^{2+}/Mg^{2+}-free balanced salt solution. Working conc. 2–3 mg ml^{-1} [111]
Papain (EC 3.4.22.2)	From *Carica papaya*. Working conc. 0.05–0.5 mg ml^{-1} (1.5–15 U ml^{-1}) in PBS or BSS. Nonspecific sulfhydryl protease. Activator: cysteine (0.5% w/v, essential), EDTA. Inhibited by PMSF, leupeptin, Hg^{2+} and other heavy metals [112]
Pronase	From *Streptomyces griseus*. Nonspecific mixture of proteolytic enzymes, including neutral and alkaline proteinases, aminopeptidases and carboxypeptidases. Often used in conjunction with other enzymes. Working conc. 0.5–2 mg ml^{-1} (3–20 U ml^{-1}). Usually used with 2 mM Ca^{2+}, thus inhibitors include EDTA. Components may be resistant to inhibition by serum. Also similar preparation known as Neutralase™ [113]

Continued

Cells

Table 9. Enzymes for tissue disaggregation, *continued*

Thermolysin	From *Bacillus thermoproteolyticus rokko* (EC 3.4.24.2). Known simply as protease in some catalogs. A metalloendopeptidase with specificity towards isoleucine, leucine, methionine and valine. Has been used to dissociate lung tissue. Inhibited by EDTA and 1,10-phenanthroline [114]
Trypsin (EC 3.4.21.4)	From bovine pancreas, though also available from porcine pancreas and human pancreas. Available as a powder or different strength solutions. Often strength indicated by 1:250 (1 part able to digest 250 parts casein under standard conditions). Can be reconstituted in either Ca^{2+}/Mg^{2+}-free PBS or EDTA/Versene. Working concentration 0.25%. Inhibited by serum, aprotinin, chicken egg white inhibitor and soya trypsin inhibitor (STI). While a popular choice for subculturing adherent cells, it was the first enzyme used for tissue dissociation. Serine protease cleaves protein at the peptide bonds involving the carboxyl group of arginine and lysine. Often used with collagenase and hyaluronidase [2, 115]

EDTA, ethylenediaminetetraacetic acid; EGTA, ethylene glycol-bis (β-aminoethyl ether) *N,N,N',N'*-tetraacetic acid; SDS, sodium dodecyl sulfate; PMSF, phenyl methyl sulfonyl chloride; DFP, diisopropyl fluorophosphate; N.F., National Formulary; USP, United States Pharmacopoeia.

Table 10. Sample tissue disaggregation protocols

Tissue pretreatment	Enzymes	Treatment	References
Ovarian tumor, minced to less than 1 mm^3	Hyaluronidase 0.1% DNase 0.01% Collagenase 2.5 U ml^{-1}	1 mM Hepes, HBSS, 2–4 h	116
Rabbit aorta, 0.5 mm pieces, McIlwain tissue chopper, washed in KRH 25°C in 0.2 mM Ca^{2+}	Elastase 80 U mg^{-1} Collagenase 450 U mg^{-1} DNase I 2800 U mg^{-1}	1 g tissue to 10 ml KRH + Ca^{2+} + Enz. 37°C, 75 min shaking. Monitor pH to 7.4	117
Meningioma tumor minced 1 mm^3	Dispase 2.4 U mg^{-1}	Ca^{2+}/Mg^{2+}-free HBSS, shake 2 h, 37°C	118
12-day Balb/c embryo, minced	Collagenase 0.1 U ml^{-1} Dispase 0.8 U ml^{-1}	PBS, 60 min 37°C, follow with 3 × PBS wash	119
Renal carcinoma, 1 mm^3 pieces	Trypsin 0.25% or Trypsin 0.25% Collagenase 0.5 mg ml^{-1} or Collagenase 0.9 mg ml^{-1} DNase 0.09 mg ml^{-1} Papain 0.09 mg ml^{-1}	10 changes of RPMI + enzymes. Incubate at 37°C, spin out free cells	120

KRH, Krebs/Ringer/Hepes; HBSS, Hanks' balanced salt solution.

Cells

Table 11. Chelating agents

Citrate
Trisodium citrate dihydrate $C_6H_5O_7Na_3.2H_2O$, mol. wt = 294.1
Can be used at any concentration between 40 μM (11.7 μg ml^{-1}) and
  1 mg ml^{-1} in Ca^{2+}/Mg^{2+}-free PBS or balanced salt solution (BSS).
  Can also be used in conjunction with trypsin [121, 122]. Has been
  largely superceded by the introduction of EDTA
Also used in ACD (acid–citrate–dextrose) buffer for preventing blood
  from clotting on exposure to the air.

> Trisodium citrate 2.2%
> Citric acid 0.8%
> Glucose 2.45%

1 part ACD added to 6.6 parts blood

EDTA
Ethylenediamine tetraacetic acid, disodium salt, dihydrate
  $[CH_2.N(CH_2.COOH).CH_2COONa]_2.2H_2O$, mol. wt = 372.24
Free acid $(HOOCCH_2)NCH_2CH_2N(CH_2COOH)_2$, mol. wt 292.2
Used at 0.02% (0.54 mM) or up to 100 mM in Ca^{2+}/Mg^{2+}-free buffer
Reversibly binds divalent cations, i.e. Ca^{2+} and Mg^{2+}

EGTA
Ethyleneglycol-bis-(β-aminoethyl)-*N,N,N',N'*-tetraacetic acid,
  mol. wt = 380.4
Used at 0.1 mM in Ca^{2+}/Mg^{2+}-free buffer
Specifically chelates calcium ions

Table 12. Attachment factors

Factor	Sub-type	Concentration
Collagen	Type 1, rat tail	5–10 μg cm^{-2}
	Type 1, calf skin	5–10 μg cm^{-2}
	Type 2, chicken	6–10 μg cm^{-2}
	Type 3, human	—
	Type 4, mouse	1–10 μg cm^{-2}
	Type 5, human placenta	1 μg cm^{-2} approx.
Fibronectin	Human plasma	1–5 μg cm^{-2} or 5 μg ml^{-1}
	Bovine plasma	as media additive
	Human cellular	
	Rat plasma	
	Human foreskin	
	Recombinant (fibroblast-like)	
Gelatin	Bovine skin	100–200 μg cm^{-2}
	Porcine skin	
Laminin	Mouse sarcoma	1–2 μg cm^{-2}
	Human placenta	
Poly-D-lysine	Synthetic, 30–70 kDa	2.5–5 μg cm^{-2}

Versene

Basically an EDTA solution made up in a phosphate buffer, this is the recipe that is used in the author's laboratory

Used as a diluent for trypsin, or as a prewash in cell passaging/subculturing to remove Ca^{2+}/Mg^{2+} and serum

	mg ml^{-1}	mM
NaCl	8.0	136.8
KCl	0.2	2.68
Na$_2$HPO$_4$	1.15	8.1
KH$_2$PO$_4$	0.2	1.47
EDTA	0.2	0.537

Adjust pH to 7.2, and either autoclave or filter sterilize

	Synthetic, 70–150 kDa	
	Synthetic, > 300 kDa	
Poly-L-lysine	Synthetic, 30–70 kDa	2.2–5 µg cm^{-2}
	Synthetic, 70–150 kDa	
	Synthetic, > 300 kDa	
Vitronectin	Human plasma	Variable,
	Rat plasma	as low as
	Bovine plasma	0.1 µg ml^{-1}

Cells

Table 13. Trade name attachment factor/gel matrix formulations

Cytopex	Imperial Laboratories. Solubilized biomatrix from EHS (Engelbreth–Holm–Swarm) murine sarcoma. Thus a major constituent is laminin. Also present are (murine) collagen type 4, heparan sulfate, growth factors such as EGF and FGF with additional TGF. Proteoglycan and plasminogen activator. Thick or thin gels can be prepared according to need. Prepared in buffered saline, thawed overnight at 4°C (it gels irreversibly at or above 22°C), and mixed. 50 µl Cytopex cm^{-2} are added to culture plates at 0°C, cells are then incubated at 37°C until gelled. Cells can be suspended in Cytopex at 200 µl cm^{-2} provided the cells do not dilute the Cytopex too much. Thin films, precoating/subbing can be achieved by diluting in chilled growth medium, adding to cover growth surface, and incubating at room temperature for 60 min.
Matrigel	Collaborative Biomedical Products. Also prepared from EHS mouse sarcoma cells and has properties apparently almost identical to those of Cytopex. In this case release of cells from the matrix by dispase is recommended (see *Table 9*).
Vitrogen 100	Collagen Corporation. Sterile solution of pepsin-solubilized bovine dermal collagen.
Cell Tak	Formulation of polyphenolic proteins extracted from *Mytilis edulis* (marine mussel). Biocompatible. Will coat most surfaces; supplied in 5% acetic acid. Used at 3.5 µg cm^{-2}.

Table 14. Strategies for the suppression of fibroblast contamination

Geneticin (G418)	100 µg ml^{-1} of active material for 48 h. Claimed to be preferentially toxic to fibroblast cells over epithelial cells [123]
Irradiated feeder cells	Plate lethally irradiated 3T3 feeder cells at one-third confluence with epidermal cells. Used for explants of epidermal keratinocytes [124]

Low calcium	Some cells types both tolerate and proliferate in low calcium medium (0.15 mM), this is not the case for fibroblast cells [125, 126]
D-Valine	Disputed claims that fibroblasts die in the absence of L-valine, while some epithelial cells can utilize D-valine as an alternative. Some manufacturers have now made available ready-made D-valine medium [127–129]
IBMX	0.5 mM in continuous culture claimed to inhibit the growth of fibroblast cells. However, as an inhibitor of cyclic adenosine monophosphate phosphodiesterase, it may be encouraging the growth of the epithelial cells, as with cholera toxin (around 10^{-10} M in epithelial culture) [130, 131]
cis-Hydroxyproline	50–200 μg ml^{-1} treatment indefinitely, slows the growth of fibroblasts (L929 and chick embryo tendon fibroblasts) [132]
Putrescine	Concentrations of 1–10 mM have been shown to inhibit fibroblast growth, whereas 1 mM has been shown to stimulate growth of 'some' epithelial cells (bronchial) [133]
Antibody-mediated cytotoxicity	Specialized treatment relies on the presence of antibody against the cell you wish to remove. Either uses toxic conjugate or the presence of complement in the serum for lysis [134,135]
Selective attachment and detachment	Scraping the culture under the microscope can allow cells of interest to be selected at the expense of the removed cells [136]. Also, 0.025% trypsin (v/v) versene has been used for the removal of fibroblasts and then used at 0.05% for passaging the epithelial cells [137]. Additionally, dispase will preferentially remove epithelial cells, leaving the fibroblasts behind. Areas of interest could also be ring cloned and cultured

Cells

Chapter 6   CULTURING CELLS

1 Daily evaluation

Cell culture is a labor-intensive business: leaving cultures for too long will cause them to overgrow and die. In contrast, too much handling can increase the risk of contamination, cause temperature stress and can interfere with cell growth. It is important for cultures to be observed daily, but also quickly and accurately.

The *medium color* can be observed easily, the *medium clarity* determined and the cells can be *observed under the inverted microscope* easily. This is easy to do, will cause the minimum of disturbance to the cultures and can yield a wealth of information (see *Table 8* for guidance with the observations made).

horizontal. To the left, and a little below, is another tap, for flushing the system.

When the glass tube is placed in contact with the cell suspension (with the vacuum pump on), turning the top tap (until either a green line appears on the size distribution display, or until a red light comes on and then goes off (different machine models)), disturbs a column of mercury. In the process of returning to equilibrium, this sucks 0.5 ml of cell suspension through the orifice in the round-bottomed glass tube. While this is taking place interference with an electrical field occurs and the cells are counted. Thus in essence, after the pump is switched on and the cells diluted in Isoton (1:20 to 1:50 v/v), they can be placed on the platform with the probe within the suspension,

2 Counting

Cultures should be counted when being subcultured and counted daily if a growth curve is to be constructed. The practical choices for counting cells in the laboratory are the humble hemocytometer and the Coulter counter. The hemocytometer is a sophisticated microscope slide, whereas the Coulter counter counts particles passing through a hole. The advantages and disadvantages of counting methods are given in *Table 1*.

2.1 Points to note on counting
Both methods are easy to do badly, and both can be hazardous to health if not performed correctly.

Coulter counter (basic models Dn or ZF)
Most Coulter counters consist of what looks like a round-bottomed glass tube (probe) dipping into a clear liquid (Isoton or equivalent) supported by a spring-loaded platform with a tap at the top. This should be set with the arms

the top tap is turned for a short time, and the cells are counted. Example:

Dilution: 0.5 ml cell suspension + 9.5 ml Isoton
Cell count: 4730
Cells ml^{-1} of original $= 4730 \times 20 \times 2 = 191\,200 = 1.9 \times 10^5$

where the multiplication by 20 is the dilution factor, and the multiplication by 2 corrects for the 0.5 ml aliquot taken up. If the count is higher than 10 000, then the sample should be diluted further.

The accuracy of the cell count is dependent on the settings of the aperture, attenuation and threshold, and for this the manual should be consulted. If you are unsure of a count given, then Coulter count a sample without cells and also repeat the cell count with a hemocytometer. If the two cell counts are different, the counter needs recalibration.

Remember that these counters can also be used for counting different cell sizes and determining size distributions.

If the waste beaker is full, the pump can suck back and the mercury can end up in the wrong part of the machine. Do not leave the machine switched on when not in use.

If the gates are not set correctly, the counter may include cell debris as cells.

Care should be taken when using cells of different sizes, i.e. compare the size of a small quiescent lymphocyte with a fibroblast after trypsinization. If the operator is not careful, the setting for one type of cell may exclude some of the population when the other cells are counted. This may lead to a tendency to underestimate cell number.

It is important to remember that inaccuracy creeps in if the cells are clumped, unhealthy, or the suspension too concentrated.

Hemocytometer
Many find difficulty with the calculations involved, and forget to account for the dilution factor when using viability stains, or fail to count the minimum number of cells (100–200).

however, we have not normally found this to be a problem. The clarity of the erythrosin B solutions means that contamination of counting buffers is detected readily. Many find that eosin is less easy to use than others, and some say that erythrosin B is less cytotoxic than trypan blue.

3 Subculturing

Suspension cells are relatively easy to subculture. After counting they can be centrifuged out of the spent medium or diluted in pre-warmed media to a lower concentration. The initial concentration will depend on the cell type in use; however, for most cell lines cultures can be initiated with $1–2 \times 10^5$ cells ml^{-1} from a maximum concentration of $8–10 \times 10^5$ cells ml^{-1}.

Adherent cells must be removed from the substratum; the ways in which this can be done are shown in *Table 3*.

In the most popular case, 'the improved Neubauer' (see *Figure 1*), the central 25 squares represents 1 mm^2, the distance between the slide and the coverslip (specially thick to avoid inaccuracy through distortion) is 0.1 mm, thus the volume represented is 0.1 mm^3. The number of cells (N) in 1 ml (1 cm^3) is the number counted (n) $\times 10^4$; if the solution had been diluted by half, then $N = n \times 2 \times 10^4$.

2.2 Stains

Stains (*Table 2*) must be made up in physiological strength buffers so that the dye solution does not cause cells to lyse (0.9% NaCl or its equivalent).

All of these stains should be mixed 1:1 with cells and incubated for 2–3 min before counting. However, times longer than 5 min should be avoided as the toxicity of the dyes may artificially reduce viability estimates [1].

Some feel that trypan blue exclusion counting should be done in the absence of serum as it stains serum proteins;

4 Cryopreservation

Cryopreservatives allow cells to be taken to low temperatures and then allow their recovery, at a later date, from that low temperature to cell growth. Dimethylsulfoxide (DMSO) is the most commonly used and works because of its ability to penetrate the cells, allowing the water content of the cells to cool without the formation of ice crystals. Glycerol can also be used.

4.1 Freezing

Warning: the penetrative ability of DMSO makes it a chemical to be treated with care as it can carry other chemicals across the membrane with it. The recipe for a freezing mixture is given in *Table 4*.

Points to note:

1. All manipulations should be done on ice or in ice-cold media.
2. Freezing mixture can be made up at double strength, and the cells resuspended in medium without serum

Culturing Cells

prior to the addition of freezing mixture. Alternatively, the mixture can be made at single strength and the pelleted cells resuspended in the mixture. Some suggest adding the freezing mixture dropwise to reduce osmotic shock.

3. Freezing mixture can be stored frozen, and filtered through a 0.22 μm filter if required. DMSO is self-sterile, although it does oxidize in air, it is best to buy small bottles and replace often.

Cell concentration: adherent, $1–2 \times 10^6$ ml^{-1}, 1 ml/vial; suspension $5–10 \times 10^6$ ml^{-1}, 1 ml/vial. These are guidelines only and may differ for specialized cell types. Cells seem to freeze best when subconfluent and during exponential growth.

Cells should be cooled at 1°C min^{-1} to at least -50°C before transfer to liquid nitrogen. Freezing plugs can be set so that the cells are in contact with the N$_2$ vapor; alternatively, polystyrene boxes (with 5–10 mm thick walls, dead-space filled with tissue paper) serve the same purpose.

5 Cloning

Cloning in cell culture is not to be confused with cloning in molecular biology. In the latter case this concerns the identification, isolation, sequencing, protein expression and manipulation, etc., of DNA. The former refers to the growth of cells in such a way that colonies of cells can be said to be the daughter cells of a single progenitor. Several cloning techniques can be used.

5.1 Semi-solid

Primarily for suspension cells, or the isolation of anchorage-independent cells.

Two methods are used to produce the semi-solid matrix: noble agar (or agarose, soft agar cloning) and methylcellulose (also known as methocel cloning). Both methods experience batch variation in their ability to support clonal cell growth.

Recently, a number of suppliers have introduced the cryobox (e.g. Nalgenes' Mr Frosty) which, when filled with isopropanol, provides a controlled cooling rate in a $-70°C$ freezer.

4.2 Thawing

Care: cryovials very often fill with liquid nitrogen, thus they can either explode, or squirt liquid nitrogen in the direction of the operator. Gloves and a safety visor MUST be worn when using liquid nitrogen.

Cells should be thawed quickly, and diluted in excess *prewarmed* medium. Some recommend the dropwise addition of medium to the cells to minimize osmotic shock. The amount of washing required is debatable, we find with some adherent cells that it is better to grow the cells overnight in what is effectively 0.5% DMSO, and change the medium later. Others suggest as many as three washes before culturing. It may thus be best to follow the guidelines for the particular cell type in question.

Soft agar
Perhaps the simplest semi-solid method (though the alternative, methocel (methylcellulose, 4000 cps) at between 0.8% and 1.3% is favored by hematologists [3]).

Stock noble agar 5%
Stock agarose 2%

Made up w/v in tissue-culture-grade water, autoclaved, kept at $60°C$ in a water bath. Agarose stock can also be stored at $4°C$, and later melted at $44°C$.

Noble agar is diluted to 0.3% in growth medium containing cells and 20% fetal calf serum (FCS); or agarose is mixed to 0.66% in medium containing cells and double-strength serum.

Remember that many cells require up to 20% FCS for growth in soft agar.

Remember to *prewarm* the medium; act promptly once the solidifying agent is to be added; allow to set at room temperature.

Culturing Cells

It is a good idea with a new cell type to try a range of cell concentrations (from 10 cells ml^{-1} to 10^4 cells ml^{-1}, depending on the cloning efficiency). Agar–cell or agarose–cell mixes can be plated out in any form, a 24-well plate to a 25 mm deep 90 mm tissue-culture dish. Feeder layers can be used as with any other type of cloning, as can conditioned medium. In addition to this, some people 'sub' their vessels, i.e. 1% agarose, or 0.5% noble agar, is placed as a thin layer in the vessel and allowed to solidify before the final cell mix is added. This is done to prevent cells falling through the matrix to the bottom of the vessel.

If necessary, the original agar can be supplemented at a later date with additional medium for feeding. Cell colonies can be picked (for instance with sterile Pasteur pipettes) and expanded after 1–4 weeks, depending on the growth rate of the cells.

5.2 Limiting dilution

Based upon the idea that if cells are diluted and plated out at a concentration such that less than one cell is present in

done with the aid of cloning rings; these can be purchased from many suppliers and can be fashioned from porcelain, steel, glass or plastic, or can be home-made from autoclave-resistant tubing.

The plates are washed in versene as normal, the sterile cloning ring placed over the colony in question (the bottom of the ring can also be dipped in sterile silicon grease to improve the seal) and trypsin added to the colony. The cells can be viewed under the inverted microscope while the trypsin is acting. Once the cells have detached, they can be removed with fresh medium to an alternative vessel. There is always the danger of over-trypsinization in these cases, so do not be tempted to go for *every* cell in the colony, it is better to get fewer, more healthy ones.

5.4 Paper circle cloning

An alternative to ring cloning (taught to DD by Dr O. Pettengill, Department of Pathology, Dartmouth Medical School, NH, USA). Using a standard office paper punch,

every well, then colonies resulting must have originated from a single cell. However, in practice it is far less simple than that, and a great deal of mathematics has been applied to the determination of the reliability of the method. Generally the ideal is to plate the cells to be cloned at a concentration of 0.3 cells per well, and leave them to form colonies. In practice, one would generally plate the cells at 300, 30, 3 and 0.3 cells per well in the absence of any data on their potential cloning efficiency. The concentration at which the cells are most likely to form colonies is thus determined empirically. Remember that some cells may not readily clone and may require specially enriched serum, a hybridoma growth supplement or, indeed, a higher than normal serum concentration, and/or feeder layer (see below).

5.3 Ring cloning

Most often used following drug selection of modified adherent cells in a cell culture dish. Often when colonies appear in a dish during a drug selection one wishes to remove and expand them as individual clones. This can be holes are punched in filter paper (Whatman 3MM), and the resultant paper circles collected; these are autoclaved and dried in the drying cabinet. Prepare the dish in the same way as for ring cloning. Instead of placing the cloning ring in position, pick up a paper circle with sterile forceps, dip it in trypsin solution, and place over the colony. The dish can either be incubated at 37°C or kept at room temperature. When the cells at the edge of the paper begin to lift, then gently agitate the circle a few times with the forceps and remove the whole paper + cells combination to a vessel containing medium. The paper circle can be removed at a later date, after it has had the time to seed cells to the new vessel.

5.5 Feeder cells

These can be of virtually any cell type, including NIH 3T3, murine peritoneal macrophages, murine spleenocytes, human embryo fibroblasts (MRC-5, IMR-90) and human peripheral blood cells. In order to be sure that the feeder cells do not interfere with the cloning and get mistaken as a clone, they are usually exposed to 20–30 Grays (2000–3000 rad) of γ-irradiation, or preincubated with 10 μg ml⁻¹ mitomycin C (stock 200 μg ml⁻¹ without serum) for 4 h, followed by extensive (3–4 times) washing with growth medium.

Remember that to be absolutely confident of the clonal nature of the cells they should be cloned twice in succession. Also, after a few generations growth the cells diverge/drift and will no longer be clonal for all their characteristics.

6 Modification of cells in culture

Cells in culture can be modified in several ways, probably the most common being mutagenesis, fusion, transfection and infection.

6.1 Mutagenesis

This has been used traditionally to produce eukaryotic cells with defined lesions in metabolic pathways. These have subsequently been used for studies on metabolic pathways

insertional mutagenesis, usually uses a disabled retroviral vector to infect the target cells and then interrupt (or otherwise interfere with) the genes of the infected cell. This can either inactivate a previously active gene [7], or activate a previously inactive gene [8]. The advantage of this method is that, in contrast to some other methods, it is possible to trace the area into which the vector has inserted itself. This ability was initially dependent on the vector containing the *sup F* tyrosine suppressor tRNA gene. This gene allows the correction of a mutation in a bacteriophage that allows its proliferation inside special bacterial hosts [9]. In practice, identifying genes in this way can be very difficult, and the more favored method to date is to use the polymerase chain reaction (PCR) to amplify the regions flanking the insertion [10].

An alternative method uses gene targeting by means of homologous recombination. However, for this to be of use you need to know (and have at least some of the clone of) the gene you wish to interfere with. This method uses transfection to deliver a cloned gene to a cell's nucleus, this can then (at very low frequency) homologously

and in many cases used as fusion partners for the generation of somatic cell hybrids. Mutagenesis comes in many forms, but can perhaps be divided up into two categories: insertional and chemical.

Chemical mutagenesis involves the use of DNA-damaging drugs such as MNNG, EMS, DMS, 8-methoxypsoralen, X-rays, γ-irradiation and UV irradiation, etc., and can cause frameshift mutations, deletions and misrepair. The precise concentration and time course of the treatment varies for each drug and cell type, and must be determined empirically.

The most important consideration in this type of work is to decide what the selection is to be for the cell of interest. This type of mutagenesis is unable to target a particular gene for mutation, thus the selection must be designed to enrich the phenotype of interest from the background of other cells. Strategies used to select variant metabolisms include thymidine kinase (TK) [4] and hypoxanthine/guanine phosphoribosyl transferase (HGPRT) [5] deficiencies (see also ref. 6 for a fuller treatment). A more recent approach, known as recombine with, and replace sections of, the resident gene. This is best done when the product of the resultant change is dominant, since it can only combine with one allele at a time [11] although modifications do now allow selection of recessive genes [12].

6.2 Fusion

Production of somatic cell hybrids by fusion of cells to form heterokaryons. The best-known type being the fusion of antibody-producing normal murine B lymphocytes with a murine myeloma cell line [13]. Two different cells can be fused together to combine their characteristics. In this particular case the immunoglobulin-secreting activity of the former with the immortalizing function of the latter. Other fusions can involve different immortalized cells with variable selectable markers.

Before you start you should:

1. If applicable for murine and rat fusions, determine your immunization schedule [14–16].

2. Select and test your screening strategy [14–16].
3. Ensure that the phenotype of your fusion partner is as it should be, e.g. 6-thioguanine-resistant myeloma cells can revert to sensitivity.

If these vital factors are not settled in advance, you may not have primed your B cells correctly; if you have, you may not be able to screen for the correct immunoglobulin (or alternative function), or you may find that the colonies you see are false positives.

Cells

Table 5 gives an abbreviated list of fusion partner cells of the most common hybridomas formed. Of course, many other cell lines can be (and are) used (including human). Nonsecretor lines are shown, which are obviously preferable when producing monoclonal antibodies. Human hybridomas have met with limited success. Limitations of these cells are witnessed by the current popularity of fusion of mouse myeloma cells with fused human peripheral blood lymphocytes in attempts to recreate the sophistication

the cells to be used, pre-fusion cytotoxicity studies on the parent cells may be informative. (For further information and a variety of alternative protocols see refs 14–16, 19.)

Feeder layers

For hybridomas, feeder layers are very often required, and are plated in advance (1–7 days) ready to receive the cell fusion products. Commercially available products attempt to replace them, and this can be very successful; however, a syngeneic or irradiated allogeneic macrophage feeder layer can serve a dual role, as a supporter of growth *and* to engulf debris. Use the cells from the peritoneal exudate at 2×10^4 per well (for 96-well plates). Usually the exudate from one mouse produces approximately 5×10^6 cells.

Spleen feeder layers can also be used for this function but can produce a lot of debris of their own when the cells start dying off. It is helpful to remove the red cells before use. Usually one mouse spleen gives 1×10^8 cells.

achieved with the murine system [17]. Heterofusions with mouse cells are particularly unstable and preferentially lose human chromosomes. Human cell lines as partners consist mostly of EBV-immortalized lymphoblastoid cell lines (LCLs) [18], although there are some myelomas available.

Most HGPRT⁻ partners should be routinely passaged in 6-thioguanine (6-TG) (20 μM, or 20 μg ml⁻¹ 8-azaguanine) every few months for 6 days (fresh 6-TG and medium after 3 days). This ensures that the cells do not give false HAT-resistant positive clones.

Fusogens

The most popular being polyethylene glycol (PEG); the two most popular molecular weights used are PEG 1500 and PEG 4000; 50% v/v (PEG melted using a 50°C water bath) with either PBS or medium without serum. Alternatively, use 50% w/v added to diluent at 37°C, aliquoted and stored at 4°C. Opinion is divided on the best molecular weight for the job, this must be decided empirically for

Media

RPMI 1640, DMEM or Iscove's (high glucose 4.5 mg ml⁻¹) have been used successfully for growth of hybridomas with 10–20% serum. Use 20% for fusion and while the fused cells are developing. Also available now are serum-free media for hybridomas, bovine colostrum supplement (instead of FCS or for reduced FCS), growth factors such as 'hybridoma fusion and cloning supplement' (Boehringer) and conditioned media from various cell lines (Sigma). Low-protein-content media can be useful when antibodies are produced, as they are more easily purified [20].

Selection

For HAT and HT selection see Section 7

6.3 Transfection

Many different methods can be used to transfect cells in culture and the list is growing all the time. Here we will present the salient points of the most common ones and introduce some that may yet become the methods of choice.

Calcium phosphate coprecipitation

Still probably the favorite method for the stable transfection of adherent cells [21,22].

- Good for adherent cells and stable transfection, although not recommended for suspension cells.
- Many modifications of the original protocols are available.
- DNA should be supercoiled, clean and free from contaminants, and can be in the presence or absence of an excess of carrier DNA (i.e. sonicated salmon sperm DNA). Some claim PEG precipitation of plasmid DNA is sufficient, others demand two rounds of purification by cesium chloride centrifugation.
- DNA should be precipitated in a mixture of calcium chloride (220–250 mM from 2–2.5 M stock) and Hepes-buffered phosphate (pH 7.05), usually prepared as concentrated stock solutions. Precipitated DNA can be added to cells in the presence or absence of culture medium and left with the cells for 1–24 hours (depending on the protocol).
- Transfection can be accompanied by glycerol shock (10–15%), DMSO (10–20%) shock or chloroquine

should be carried out with the cells you are using (stock, 10–50 mg ml^{-1}, can be filter sterilized and stored at $-20°C$; range in culture 100–400 μg ml^{-1} for up to 4 h, optimum often around 200 μg ml^{-1}), and the optimum DNA concentration should also be determined (1–10 μg ml^{-1} in medium).

- DNA is thought to bind to the DEAE-dextran which, in turn, sticks to the cells' surface. This transfection is usually accompanied by a DMSO or glycerol shock.
- Do not add high-concentration DNA directly to media containing serum and DEAE-dextran, the DNA will tend to clump; instead, it can be added slowly (with constant shaking) to warm, high-concentration DEAE-dextran. The other alternative is to add the DNA to medium without serum and then carefully add the DEAE-dextran.
- Once again, the adherent cell starting number should be judged in the same way as the calcium phosphate technique, although some prefer to remove their cells from the substratum and transfect the cells in suspension [25].

$(100 \ \mu g \ ml^{-1})$ treatment after the DNA precipitation. These treatments are all cytotoxic and thus the duration and concentration used should be determined empirically. Some say efficiency is improved if the phosphate buffer is made with BES ($N'N$-bis(2-hydroxyethyl)-2-(aminoethanesulfonic acid, mol. wt 213.3) as it is claimed that the precipitate forms more slowly and is thus finer; this method is often used in conjunction with culture medium in overnight precipitations on to the cells [23].

DEAE-dextran

A favored method for highly efficient transient transfections [24]. Often used on Cos-1 cells with plasmids containing the SV40 origin of replication, which can replicate episomally in these cells. Some find that this is also better for some suspension cells that can be difficult to transfect with the other available methods. Many workers also find this easier to use and more reproducible than calcium phosphate.

- DEAE-dextran (mol. wt 5×10^5 up to 2×10^6) is a long-chain polycation, toxic to some cells. Pilot incubations

- Optimization and control for DNA transfection efficiency is important for transient transfections, since these are often used to compare different plasmid constructs for effects such as transcription.
- Many protocols are available, and once again kits can be purchased from suppliers such as CP Laboratories, Pharmacia, Promega and Stratagene.

Lipofection

Originally this method required the worker to make up his or her own liposomes [26] for liposome-mediated gene transfer. These could be difficult to make, the lipid had to be stored under argon to prevent oxidation, and the process was fraught with problems and *very* unreliable. However, in recent years it has become a very popular technique for both adherent and suspension cells, with a number of companies now selling lipofection kits, for example: Transfectam (Promega); DOTAP (Boehringer Mannheim); Lipofectamine™, Lipofectin™ and LipofectACE™ (Gibco-BRL).

Culturing Cells

Originally it was thought that the DNA is transferred inside the lipid envelope, and when this sticks to the cells the lipid fuses with the membrane and the DNA enters the cell. It now appears that the DNA does not need to be inside a lipid envelope, but merely associated with it.

Lipofection appears to be better for transient expression rather than stable integration.

The reagents are usually nontoxic; thus if few cells die and there is a reasonable efficiency, relatively high expression levels can be obtained.

It appears that, up to a maximum of about 8 h, the longer the cells are in contact with the lipid the better the transfection.

As with most transfections, to obtain the best results it is necessary to optimize the method for the cells of your choice, this includes the concentration of the DNA, the lipid and the length of exposure.

mimic the conditions found inside the cells; however, now the emphasis has changed and this no longer seems so crucial.

The main drawback to this method is the expense of the pulse equipment and the disposable cuvettes (although some workers clean and reuse their cuvettes).

It is also not a particularly good method for obtaining stable integrations. On the other hand, it is the only method that will successfully introduce DNA into some cell types.

Many say that optimum transfection is more easily achieved using linear DNA, although this does cost quite a lot, in restriction enzymes, to prepare. Also remember not to interrupt the gene of interest with the digest.

Generally, the required DNA concentrations are suggested to be close to 10–40 µg per 10^7 cells.

The pulse delivered is usually a single pulse at approximately 1.5 kV and 125 µF; remember that some cuvettes have

The lipofection kits are simple to use and often give good results and therefore should be tried, even though some cells may not respond well to this method. Despite the fact that the kits can be very expensive, we advise that you avoid trying to make your own liposomes, we have had *very* variable results using this route.

Electroporation

This is the favorite method for those who have found their suspension cells difficult to transfect [27,28], and can be used with either adherent or suspension cells. The general principle is that the cells are suspended in a buffer containing DNA and a voltage is applied to the buffer and the cells, as a result of which the membrane of the cells becomes 'peppered' with holes and/or the membrane loses its polarity. The upshot of this transient effect is that the DNA is able to enter the cells while their 'defenses' are down. Opinion is constantly changing about the buffer to use for these electroporation experiments. In the past, recommendations have included growth medium with and without serum, PBS and mannitol. Originally, in order to reduce toxicity, workers went to some lengths to

different pathlengths (i.e. the distance between the plates), and that if the cuvette is not full, it will discharge through a smaller cross-section of the plate, i.e. only that covered by the medium.

Opinion is divided, but the consensus is that the pulse should be delivered at 0°C, and the cells kept for 10 min at 0°C before they are further manipulated. They tend to be *very* delicate at this stage and any dilution that needs to be done should be performed *very* carefully. Often a survival of 20–30% is found 24 h after the transfection. In general it seems that the lower the toxicity, the lower the transfection efficiency.

General points on transfection

It is generally agreed that the health of the cell prior to transfection is one of the most important keys to success. This, coupled with the purity of the DNA, can make all the difference to a successful transfection. Several other methods have been used, i.e. high-velocity microprojectiles [29,30], scrape loading [2], protoplast fusion [31] and microinjection [32], all of which have their relative merits and demerits.

Culturing Cells

However, many felt that the utilization of the infective ability of viruses was the best route to increasing the scope and efficiency of this technique.

6.4 Infection

The two most common infectious agents in use at the present time to modify cells in culture are the Epstein–Barr virus (EBV) and murine amphotropic retroviral vectors.

EBV infection

This is severely limited in its appeal as only the genes that the EBV carries are introduced into the cell. Also the type of cells that can be infected are limited to B lymphocytes (or, at least, cells expressing the complement receptor CR2). Despite this, the virus has for a long time been used for the immortalization of B lymphocytes to give B-LCLs (B lymphoblastoid cell lines).

LIVE INFECTIOUS EBV CAN INFECT THE LAB WORKER. All work must be done in a class II recirculating virus supernatant. We would prefer not to include the added complication of the phorbol ester in method 1. After the cells have been allowed to settle, the supernatant can be spun at 400 g for 10 min, filtered through a 0.45 μm filter, aliquoted into cryovials (1 ml per vial) and stored at $-70°C$.

Cyclosporin A (0.5–2 μg ml^{-1}) can be used when infecting adult cells, since any T cells present need to be inhibited.

Always include noninfective control wells, as the cells have a tendency to clump even in the absence of immortalization.

Retroviral infection

Use of disabled murine retroviral vectors is one of the easiest ways to transfect cells. The DNA of interest can be cloned into the vector and, following a long series of manipulations, a cell line can be created that leaks infectious (though with a disabled life-cycle) retrovirus carrying the gene of interest. These vectors will infect human and mouse cells, both suspension and adherent cells [33].

hood; all plasticware should be soaked in a strong oxidizing agent after use. If the hood has a UV light, it should be used after the manipulations have carried out. See also the Appendix.

Infectious virus can be obtained from the B95-8 marmoset cell line (see Chapter 5), the B95-8 refers to the strain of virus as much as to the cell line. A similar such strain, P3-HR1, although infectious, does not immortalize.

Infective virus is leaked into the medium by B95-8 cells, and can be increased by stressing the cells. This is done in one of a number of ways, as follows. When the cells have reached high concentration ($1–2 \times 10^6$ cells ml^{-1}), then either:

1. Incubate with 10 ng ml^{-1} 12-O-tetradecanoylphorbol 13-acetate for 24 h at 37°C;
2. Incubate at 33°C for 48 h;
3. Incubate at room temperature for 24 h; or
4. Incubate at 4°C overnight.

All of these will result in the production of live infectious

- The disadvantage of this type of infection is that the DNA of interest requires a great deal of preparatory work before the 'producer cells' (those that produce the infective retrovirus) are ready.
- Infections can be performed with cell supernatant, live producer cells, or dead, i.e. mitomycin C or γ-irradiated, cells. In addition, the supernatant can be passed through a 0.45 μm filter (this causes a significant, though not prohibitive, drop in viral titer).
- If you intend to use live producer cells, remember that any selection process under which you place the target cells is likely to select for the contaminating producer cells as well.
- In order to increase the infectivity of the virus, some workers infect in the presence of 8 μg ml^{-1} polybrene or 5 μg ml^{-1} protamine sulfate, with or without 0.2% DMSO.
- Do not grow the producer cells for too long as the titer tends to drop. Additionally, cells that have been transfected by this means tend to lose the expression of the gene after prolonged growth *in vitro*.
- Remember that packaging cells have also been transfected with drug-resistance genes, this limits those that

Culturing Cells

can be used in the vector. For instance, hygromycin B vector cannot be used with GP + envAM12 cells.

- The size of the gene that can be transfected by this method is around 9 kb, although in practice the titer of the virus with an insert of that size would be very low. It is best to stick to less than 6 kb at the moment.
- Potentially, all cells carrying retroviral vectors can produce RCRs (replication competent retrovirus) if they become superinfected with a competent retrovirus.

Adenovirus

This relatively recent innovation is one of the few methods capable of successfully transfecting cells that are not actively proliferating; it has huge potential for transfection practitioners because of its potentially high efficiency [34]. The main problem is that the virus is live and the gene of interest must once again be engineered into a new vector (in this case one that allows recombination with the adenovirus genome while the two are inside a permissive cell line). For this reason reports on the efficacy of adenovirus conjugates with DNA and polylysine–transferrin as a transfection system are

between new data that refer to units and old data that are in $\mu g \ ml^{-1}$ (conversion rates seem to be approximately 1000 U mg^{-1} hygromycin). Store stock for long periods at $-20°C$, although it appears to be stable at $4°C$. The toxicity of hygromycin B appears to be slightly density dependent, i.e. at high cell concentrations more drug is required to kill the cells. This effect may be the result of some limited endogenous phosphotransferase activity detoxifying the medium.

7.2 Geneticin (G418)

Blocks protein synthesis by interfering with ribosomal function. Resistance gene functions by inactivating and thus detoxifying the drug by phosphorylation. Cells have variable resistance and can take up to 1 week to die, adherent cells seem to be more sensitive. Beware of the concentration advice given: in most preparations the active G418 is around 50%, therefore '1 mg ml^{-1}' would be approximately 500 $\mu g \ ml^{-1}$ of *active* G418.

exciting considerable interest [35, 36]. We would not be surprised to see kits comprising transferrin–polylysine and inactivated adenovirus commercially available very soon.

7 Selection of modified cells in culture

Many of the modifications to cells in culture are rather inefficient, thus in order to produce an enriched population of modified cells there must usually be a selection system that will kill (or inhibit the growth of) the unmodified cells while allowing (or even encouraging) the growth of the modified cells. Methods used are shown in *Table 6*.

7.1 Hygromycin B

Inhibits protein synthesis by disrupting translocation and promoting mistranslation. Suspension cells seem resistant to the high end of the concentration range, whereas most adherent cells are killed in 100–200 µg ml^{-1} in 4–5 days. Hygromycin B now comes with activity data, which have not previously been available; there may be some confusion

7.3 Puromycin

Mimics endogenous aminoacyl-tRNA and causes premature termination of the polypeptide during protein synthesis. Acts very quickly and can kill 99% of cells within 48 h; the resistance gene gives very effective protection. Little apparent difference has been seen between the sensitivity of adherent and suspension cells in our hands.

7.4 L-Histidinol

Acts in two separate but related ways. L-Histidinol is toxic to cells, thus the His D gene detoxifies the L-histidinol to L-histidine. In medium lacking the essential amino acid L-histidine, the product of the L-histidinol dehydrogenase is thus a substrate for the growth of the cells. This dual action means that optimal conditions for initial selection will not necessarily be the same for long-term growth. The concentration required to kill unmodified cells may be more than is required for the growth of the transfectants.

Culturing Cells

7.5 Mycophenolic acid

Mycophenolic acid blocks inosine monophosphate (IMP) to xanthine monophosphate (XMP), and aminopterin blocks the biosynthesis of IMP. The cells are therefore dependent on the presence of hypoxanthine/guanine phosphoribosyl transferase enzyme in the presence of xanthine. Hypoxanthine and thymidine are also required if aminopterin is used. This selection can work without aminopterin but is less stringent.

The main benefit of this selection is the use of cells that still have a functional endogenous HGPRT enzyme.

7.6 Bleomycin/phleomycin/zeomycin

These are all basically the same drug, which is believed to kill cells as a result of its ability to intercalate into DNA and degrade it. This can be quite a slow process, in some cases taking longer than G418 to kill untransfected cells.

The above represent the most often used selections, with dominant selectable markers that can be used in cells that are no aminopterin) for 1–2 weeks. Maintenance of HGPRT deficiency is achieved with 6-thioguanine (2×10^{-5} M) or 8-azaguanine (20 µg ml^{-1}) [45].

This type of negative selection with cytotoxic prodrugs has been a popular choice for the selection of biochemically deficient cells. One interesting lateral view of this was the use of the toxic thymidine analog and prodrug ganciclovir for the selection against the presence of the herpes simplex thymidine kinase gene (HSV-TK). This relied upon the relative specificity of the endogenous thymidine kinase compared with the viral homolog. Cells without HSV-TK were completely resistant to 5 µM ganciclovir, whereas it was toxic to 100% of those with HSV-TK [12].

8 Problems with cells in culture

A list of the things that could possibly go wrong would fill a book many times the size of this; *Table 8* gives a brief résumé of the most common problems and, where possible, some solutions.

otherwise essentially normal. There are also selections that can be performed in slightly different ways, using previously modified cells, such as methotrexate amplification of dihydrofolate reductase genes [44]. Finally, one of the most important selections used at the moment is HAT selection [43] (*Table 7*).

7.7 HAT selection

In this case both the thymidine kinase (TK) gene and the hypoxanthine/guanine phosphoribosyl transferase (HGPRT) gene must be present for cells to grow in the presence of HAT (hypoxanthine, aminopterin, thymidine). As before, the aminopterin blocks the *de novo* synthesis of nucleotides, making the cells dependent on the salvage pathway enzymes TK and HGPRT. This strategy is most often used in the selection of B-cell hybridoma cells, but can also be used to select for the daughter cells of any fusion between TK⁻ and HGPRT⁻ cells. Remember that when the HAT selection is relaxed, the aminopterin still in the cells can inhibit the growth, thus cells are usually grown in HT (same as HAT but with

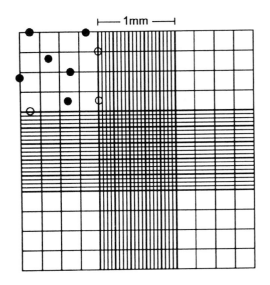

Figure 1. Scheme of an improved Neubauer chamber. Thick lines indicate triple grid. Circles in the upper left 1 × 1 mm square indicate objects to be included (filled) and objects to be excluded (open). Reproduced from C.P. Rubbi (1994) *Light Microscopy: Essential Data.*

Culturing Cells

Table 1. The advantages and disadvantages of different counting methods

Hemocytometer	Coulter counter
Laborious	Quick
Accurate	Very accurate
Cheap	Expensive
No maintenance	Needs maintenance and recalibration
Easy to use	Easy to break
Viability assessment included in count	No idea of viability
No running costs	Requires special diluent
Simple concept	Less simple concept
Visualization of the cells	No visualization of cells
Cell clumps can be noted	Cell clumps not immediately obvious

Table 2. Routine viability stains for microscopy

Trypan blue	0.2% in PBS or HBSS	Viable cells do not stain, dead cells will either not appear, or stain blue, depending on the type of microscope used
Eosin (sodium salt)	0.5% in 0.9% NaCl	Viable cells do not stain, the nuclei of dead cells stain black
Nigrosin	0.2% in 0.9% NaCl (stock 2% in H_2O)	Viable cells do not stain, dead cells stain black
Erythrosin B	0.1–0.4% in PBS[a]	Viable cells do not stain, dead cells stain black

[a] pH 7.2–7.4 with 1 M NaOH.
HBSS, Hanks' balanced salt solution; PBS, phosphate-buffered saline.

Table 3. Removal of adherent cells from substratum

Cell scraping	The use of rubber policemen is widespread, and sterile cell scrapers are readily available, although quite expensive. The advantage of this method is that proteins that you may wish to detect on the cell surface are not digested away. Disadvantages include cell clumping (makes counting difficult), reduced viability and uptake of undesirable media components resulting from membrane damage [2]. This method is not usually used for routine subculture
Versene	EDTA solutions (versene is another name for an EDTA solution) are often used, along with trypsin (see Chapter 5, *Table 11* for formulation). If the cells are sufficiently confluent then versene alone will suffice, either at room temperature for 10–15 minutes or, more commonly, after 10–15 minutes at 4–10°C. This treatment will remove most adherent cells from the substratum. The advantage of this method is that surface proteins that may be of interest are not digested. Disadvantages include: very labor intensive, sometimes needs quite vigorous pipetting, cells may need to be spun out of the versene, some workers do not like exposing their cells to EDTA. Not recommended for routine subculture.
Dispase	See Chapter 5, *Table 9*. Often epithelial cells respond better to dispase than to trypsin; also, trypsin can be very poor at releasing epithelial cells from the substratum. Disadvantages include the time required (anything up to an hour or two) and cells must be spun out of the medium as dispase is not inhibited by serum. However, since it is not inhibited by serum, dispase can be added directly to the culture initially.
Trypsin	The favorite subculturing agent, quick and reasonably gentle, although the treatment can be cytotoxic if left for too long, and over-trypsinization is a common cause of subculture problems. Most often used in conjunction with a versene (or 0.02% EDTA) prewash followed by 0.25% trypsin w/v in EDTA. Incubated either cold or at 37°C until the cells begin to detach, follow with (at least) a tenfold dilution in serum-containing media. Serum-free culture must either be rapidly centrifuged out, or trypsin inhibitors such as soybean trypsin inhibitor must be used.

Culturing Cells

Table 4. Freezing mixture; final concentration of constituents

	Range	Most common examples	
DMSO	10–15%	10%	10%
Serum	10–90%	20%	90%
Medium (antibiotic free)	0–80%	70%	—

Table 5. Popular cell fusion partners

Name	Strain	Additional information
Mouse		
P3/X63-Ag8-6.5.3.	Balb/c	Nonsynthesizing nonsecreting murine myeloma, HGPRT⁻ B-cell partner
Sp2/0-Ag.14	Balb/c	Myeloma/lymphoid hybrid, nonsynthesizing nonsecretor, HGPRT⁻ B-cell partner
BW/5147	AKR	Thymoma, HGPRT⁻, fusion partner for murine T cells for T-cell hybridomas
Rat		
YB2/3.0Ag20 (YB2)	Lou/C	Nonsecretor, HGPRT⁻, B-cell partner
YB2/0	Lou/C/A0	Hybrid cells of Lou/C YB2 and A0 spleenocyte, HGPRT⁻, B-cell fusion partner

Table 7. HAT selection

Drug name	Gene/s selected	Recommended concentration
HAT	HGPRT	Hypoxanthine (0.1 mM)
	TK	Thymidine (16 μM)
		Aminopterin (0.4 μM)

Table 6. Positive selection of dominant genes

Drug name	Gene selected	Recommended concentration
Hygromycin B	Hygromycin B phosphotransferase (HPH; (Hygro^R))	100–400 µg ml^{-1} Stock 100 mg ml^{-1} in water (4°C) [37]
Geneticin G418	Neomycin phosphotransferase (Neo^R)	100–750 µg ml^{-1} 50–400 µg ml^{-1} Stock 100 mg ml^{-1} in water (4°C) [38]
Puromycin	Puromycin acetyl/transferase (Puro^R)	1–5 µg ml^{-1} Stock 5 mg ml^{-1} in water (−20°C) [39]
L-Histidinol	Histidinol dehydrogenase (His D)	0.1–3 mM Stock 20 mM in water (−20°C) [40]
Mycophenolic acid	Xanthine/guanine phosphoribosyl transferase (gpt or XGPRT)	Mycophenolic acid (78 µM) Xanthine (1.44 mM) Thymidine (48.6 mM) Hypoxanthine (1 mM) Aminopterin (4.5 µM) Glycine (133.2 µM) [41]
Bleomycin/ phleomycin/ zeomycin	Tn5 ble Sa ble Sh ble	Phleomycin (0.1–50 µg ml^{-1}) Zeomycin (100–500 µg ml^{-1}) [42,43]

Culturing Cells

Table 8. Problems with cells in culture

Appearance	Probable cause
Macroscopic	
Cloudy and yellow colored	Contamination, most likely candidate is either yeast or bacteria; most effective solution is to discard affected cultures and thaw new cells (see also Microscopic appearance)
Yellow colored and clear	1. Confluent/overgrown cells, best to discard, but valuable cells may recover if split into fresh medium and cared for. However, cell metabolism may be affected in unknown ways
	2. pH of medium too acid (low pH), but not as a result of cell metabolism – most likely due to too high a level of CO_2 in the incubator, should be 5–7%, depending on the medium in use
Bright red/ purple colored	pH of the medium too alkaline (too high)
	1. Most likely cause is too little CO_2 in the incubator – check supply for leaks and/or empty cylinder
	2. If this has happened to a limited number of cultures, it may also be the result of caps being too tight on the flasks
Floating 'cotton wool'	Fungus; this need not actually turn the medium yellow if in the early stages. Best option is to discard immediately. It is possible that the infection will not spread if the fruiting bodies have not yet matured. Also inspect closely the inside of the incubator as this type of contamination is often spread from a medium spill encouraging the growth of a contaminant (see also Microscopic appearance).
Microscopic	
Cells look 'ill'[a]	1. Mycoplasma is always a prime suspect, but it is not *always* the culprit. Commercial detection kits are available from many suppliers, in addition the ECACC and other institutes can carry out tests on your cell lines. One alternative that we do not recommend, although it can be used, is to see if the culture looks better in the presence of known inhibitors of mycoplasma growth, such as ciprofloxacin, ICN mycoplasma removal agent, Boehringer Mannheim BM cyclin. If in doubt, it is best to discard the culture
	2. Bacteria: a chronic, low-level infection that is controlled but not eliminated by antibiotics can release toxins into the medium. Sometimes this can be confirmed by growth of the cells in antibiotic-free medium; if the bacteria are there, they may show themselves

	3. Batch change in media components. Fastidious cells may be sensitive to changes in the medium, such as new fetal calf serum, old glutamine, poor water or missing supplements (i.e. nonessential amino acids)
	4. Adherent cells over-trypsinized or (adherent and suspension) cells subcultured to a lower than recommended cell concentration
Filaments in culture	Fungus, what you are seeing are the hyphae. Discard before they develop to a macroscopic scale
String of 'pearls'	Yeast often grow in strings; 5–10 times smaller than the average cell. Sometimes they grow in colonies if the culture has not been disturbed. Discard the culture
Cell-associated debris	Sometimes cultures look unhappy for no discernable reason, tending to clump, having more than the average amount of debris and sometimes very small specks that may appear motile (beware – it could be normal debris under brownian motion). Chronic low-level bacterial infection is very difficult to spot. If it is bacteria, it will most likely take over the culture. If in doubt discard, although we have, once in a while, succeeded in clearing a culture with a change to different antibiotics, but this is not recommended

Abnormal growth characteristics

Slow growth   Often wrongly attributed to mycoplasma, can be caused by any one of hundreds of reasons. Occasionally cultures will cease normal growth for no apparent reason. It is, however, worth a quick check on some of the following:

1. The age of the medium, coupled with storage conditions;
2. Glutamine – age, storage, time at $1 \times$ in final medium;
3. Serum – new batch?, wrong kind (i.e. NCS), incorrect concentration;
4. See also Macroscopic appearance, contamination;
5. Temperature of incubator, or CO_2 conc.;
6. Humidity – check that water covers the bottom of the incubator;
7. Plasticware – new?, wrong type (i.e. microbiological grade rather than tissue-culture grade);
8. Rough treatment of cells – see also Cells look 'ill', over-trypsinization – and splitting of an overgrown culture, or, conversely, cell density too low (see also Cells look 'ill')

Continued

Culturing Cells

Table 8. Problems with cells in culture, *continued*

Faster than normal growth	Not so much a problem as a well-known cell culture phenomenon – many cells in culture now grow much faster than their original description suggests, in many cases these cells have, over a period of time, reached a high passage number and have become more adapted to growth *in vitro*. Additionally, with some adherent cells there can be a tendency towards anchorage independence, especially if cells are often grown to a consistently high cell density. If this is a problem, go back to an early passage number and create a mini cell bank of your own in liquid nitrogen to preserve the characteristics that you need. For instance, remember that cells such as NIH3T3 have been very carefully cultured to maintain the originally described contact inhibition – it only takes a few rounds of poor culture for that to begin to become compromised

[a]'Ill' covers a multitude of possible appearances – suspension cells that usually look well-rounded and plump can begin to look semi-adherent, ragged, granular, dark (rather than with a bright ring around them), growing atypically and with attendant debris. Adherent cells can have all of the above appearances and, in addition, they can begin to lift from the substratum.
NCS, newborn calf serum.

Chapter 7 MANUFACTURERS AND SUPPLIERS

American Type Culture Collection (ATCC), Sales and Marketing Department, 12301 Park Lawn Drive, Rockville, MD 20852, USA.
Tel 301 881 2600.
Fax 301 816 4367.

Amgen Ltd, 240 Cambridge Science Park, Milton Road, Cambridge, CB4 4WD, UK.
Tel 01223 420305.
Fax 01223 423049.

Applied Immune Sciences Inc. (AIS) (USA), 200 Constitution Drive, Menlo Park, CA 94025–1109, USA.

UK distributors: Techgen International Ltd, Suite 8, 50 Sullivan Road, London, SW6 3DX, UK.

Becton-Dickinson (UK), Between Towns Road, Cowley, Oxford, OX4 3LY, UK.
Tel 01865 777722.
Fax 01865 717313.

Becton-Dickinson (USA), Clay Adams Division, 299 Webro Road, Parsippany, NJ 07054, USA.
Tel 201 847 6800.
Fax 201 847 6475.

Bibby-Sterilin, Tilling Drive, Stone, Staffs, ST15 0SA, UK.
Tel 01785 812121.
Fax 01785 811064.

Bio-Rad Laboratories Ltd, Bio-Rad House, Maylands Avenue, Hemel Hempstead, Herts, HP2 7TD, UK.
Tel 01442 232552.
Fax 01442 259118.

Boehringer Mannheim UK, (Diagnostics and Biochemicals) Ltd, Bell Lane, Lewes, East Sussex, BN7 1LG, UK.
Tel 01273 480444.
Fax 01273 480266.

Boehringer Mannheim GmbH, Biochimica, PO Box 31 01 20, D-6800 Mannheim 31, Germany.
Tel 0621 7590.
Fax 0621 7594004.

Boehringer Mannheim (USA), 9115 Hague Road, PO Box 50414, Indianapolis, IN 46250-0414, USA.
Tel 800 262 1640.
Fax 317 576 2754.

Calbiochem-Novabiochem (UK) Ltd, Boulevard Industrial Park, Padge Road, Beeston, Nottingham, NG9 2JR, UK.
Tel 0115 930840.
Fax 0115 930951.

Collaborative Biomedical Products
Part of Becton-Dickinson Labware.

Collagen Corporation
Obtained through Imperial Laboratories Ltd.

Corning Inc., Science Products Division, HP-AB-03, Corning, NY 14831, USA.
Tel 607 974 4667.
Fax 607 974 7919.

CP Laboratories, PO Box 22, Bishop's Stortford, Herts, CM23 3DX, UK.
Tel 01279 758200.
Fax 01279 755785.

Dynal (UK) Ltd, 10 Thursby Road, Croft Business Park, Bromborough, Wirral L62 3PW, UK.
Tel 0151 346 1234.
Fax 0151 346 1223.

Cambridge Bioscience, 25 Signet Road, Stourbridge Common Business Centre, Swann's Road, Cambridge, CB5 8LA, UK.
Tel 01223 316855.
Fax 01223 60732.

Cedarlane Laboratories Ltd, 5516–8th Line, R.R.2 Hornby, Ontario L0P 1E0, Canada.
Tel 905 878 8891.
Fax 905 878 7800.

UK distributors: VH Bio Ltd, PO Box 7, Gosforth, Newcastle-upon-Tyne, NE3 4DB, UK.
Tel 0191 492 0022.
Fax 0191 410 0916.

Clonetics Corp. Inc.
Through Tissue Culture Services.

Clontech
Through Cambridge Bioscience.

Dynal Inc., 5 Delaware Drive, Lake Success, NY 14601, USA.
Tel 800 638 9416.
Fax 516 326 3298.

Dynal International, PO Box 158, Skøyen, N-0212, Oslo, Norway.
Tel 472 52 9450.
Fax 472 50 7015.

European Collection of Animal Cell Cultures (ECACC), PHLS Centre for Applied Microbiology and Research, Porton Down, Salisbury, SP4 0JG, UK.
Tel 01980 610391.
Fax 01980 611315.

Genzyme Corporation, One Kendall Square, Cambridge, MA 02139-1562, USA.
Tel 617 252 7500.
Fax 617 252 7759.

Manufacturers and Suppliers

Genzyme Diagnostics, 50 Gibson Drive, Kings Hill, West Malling, Kent, ME19 6HG, UK.
Tel 0800 373415.
Fax 01732 220024/5.

Genzyme (UK) Ltd, 37 Hollands Road, Haverhill, Suffolk, CB9 8PU, UK.
Tel 01440 703522.
Fax 01440 707783.

Gibco-BRL Life Technologies Ltd (UK), PO Box 35, Trident House, Renfrew Road, Paisley, PA3 4EF, UK.
Tel 0141 814 6100.
Fax 0141 887 1167.

Gibco-BRL Life Technologies Inc. (USA), 8400 Helgerman Court, Gaithersburg, MD 20877, USA.
Tel 301 840 8000.
Fax 301 670 8539.

Millipore UK Ltd, The Boulevard, Blackmoor Lane, Watford, WD1 8YW, UK.
Tel 01923 816375.
Fax 01923 818297.

Millipore Corp. (USA), 80 Ashby Road, Bedford, MA 01730, USA.
Tel 617 275 9200.
Fax 617 271 0290.

Nalgene, Nalge (Europe) Ltd, Foxwood Court, Rotherwas, Hereford, HR2 6JQ, UK.
Tel 01432 263933.
Fax 01432 351923.

NIBSC (National Institute for Biological Standards and Control), Blanche Lane, South Mimms, Potters Bar, Herts, EN6 3QG, UK.
Tel 01707 654753.
Fax 01707 646730.

ICN Biomedicals Ltd (European HQ), Thames Park Business Centre, Wenman Road, Thame, Oxfordshire, OX9 3XA, UK.
Tel 01844 213366.
Fax 01844 213399.

ICN Biomedicals Inc. (USA), 3300 Hyland Avenue, Costa Mesa, CA 92626, USA.
Tel 800 854 0530.
Fax 800 334 6999.

Imperial Laboratories (Europe) Ltd, West Portway, Andover, Hants, SP10 3LF, UK.
Tel 01264 333311.
Fax 01264 332412.

Invitrogen
Through R&D Systems.

Jencon (Scientific) Ltd, Cherrycourt Way, Industrial Estate, Stanbridge Road, Leighton Buzzard, Beds, LU7 8UA, UK.
Tel 01525 372010.
Fax 01525 379547.

Nycomed UK Ltd, Nycomed House, 2111 Coventry Road, Sheldon, Birmingham, B26 3EA, UK.
Tel 0121 742 2444.
Fax 0121 742 8782.

Nycomed AS, Pharma Diagnostics, Sandakervn 64, PO Box 4284, Torshov, N-0401, Oslo 4, Norway.
Tel 472 226350.
Fax 472 712535.

PeproTech EC Ltd, 10 FitzGeorge Avenue, London, W14 0SN, UK.
Tel 0171 603 8288.
Fax 0171 603 8233.

PeproTech, Inc., Princeton Business Park, 5 Cresent Avenue, G2, PO Box 275, Rocky Hill, NJ 08553, USA.
Tel 609 497 0253.
Fax 609 497 0321.

Pharmacia Biotech Ltd, 23 Grosvenor Road, St Albans, Herts, AL1 3AW, UK.
Tel 01727 814000.
Fax 01727 814001.

Pharmingen, 10975 Torreyana Road, San Diego, CA 92121, USA.
Tel 619 792 5730.
Fax 619 792 5238.

UK distributors: Cambridge Bioscience.

Promega Ltd (UK), Epsilon House, Enterprise Road, Chilworth Research Centre, Southampton, SO1 7NS, UK.
Tel 01703 760225.
Fax 01703 767014.

Promega Corporation (USA), 2800 Woods Hollow Road, Madison, WI 53711-5399, USA.
Tel 608 274 4330.
Fax 608 273 6967.

Serotec Ltd, 22 Bankside, Station Approach, Kidlington, Oxford, OX5 1BR, UK.
Tel 01865 379941.
Fax 01865 373899.

Sigma Chemical Company (UK), Fancy Road, Poole, Dorset, BH17 7NH, UK.
Tel 0800 447788.
Fax 01202 715460.

Sigma Chemical Company (USA), 3050 Spruce Street, PO Box 14508, St Louis, MO 63178, USA.
Tel 314 771 5750.
Fax 800 325 5052.

Stratagene Ltd, 140 Cambridge Innnovation Centre, Cambridge Science Park/Milton Road, Cambridge, CB4 4GF, UK.
Tel 01223 420995.
Fax 01223 420234.

R&D Systems Europe Ltd, 4–10 The Quadrant, Barton Lane, Abingdon, OX14 3YS, UK.
Tel 01235 529449.
Fax 01235 533420.

R&D Systems Inc. 614 McKinley Place NE, Minneapolis, MN 55413, USA.
Tel 612 379 2956, 800 343 7475.
Fax 612 379 6580.

Stratagene Cloning Systems, 11099 North Torrey Pines Road, La Jolla, CA 92037, USA.
Tel 619 535 5400.
Fax 619 558 0947.

Tissue Culture Services, TCS Biologicals Limited, Botolph Claydon, Buckingham, MK18 2LR, UK.
Tel 01296 714071.
Fax 01296 714806.

REFERENCES

Chapter 2

1. Reitzer, L.J., Wice, B.M. and Kennel, D. (1979) *J. Biol. Chem.* **254,** 2669.
2. Itagaki, A. and Kimura, G. (1974) *Exp. Cell Res.* **83,** 351.
3. Swim, H.E. and Parker, R.F. (1958) *J. Biophys. Biochem. Cytol.* **4,** 525.
4. Sigma Chemical Company (1994) *Sigma Cell Culture,* Catalog and price list, pp. 185. Sigma, St Louis.
5. Eagle, H. (1959) *Science* **130,** 432.
6. Eagle, H. (1955) *J. Biol. Chem.* **214,** 839.
7. Eagle, H. (1955) *Science* **122,** 501.
8. Dulbecco, R. and Freeman, G. (1959) *Virology* **8,** 396.
9. Iscove, N.N. and Melchers, F. (1978) *J. Exp. Med.* **147,** 923.
10. Morgan, J.F., Morton, H.J. and Parker, R.C. (1950) *Proc. Soc. Exp. Biol. Med.* **73,** 1.
11. McCoy, T.A., Maxwell, M. and Kruse, P.F. (1959) *Proc. Soc. Exp. Biol. Med.* **100,** 115.

27. Dulbecco, R. and Vogt, M. (1954) *J. Exp. Med.* **99,** 167.
28. Ames III, A. and Nesbett, F.B. (1981) *J. Neurochem.* **37,** 867.

Chapter 3

1. Barnes, D. and Sato, G. (1980) *Anal. Biochem.* **102,** 255.
2. Macaig, T., Nemore, R.E., Weinstein, R. and Gilchrest, B.A. (1981) *Science* **211,** 1452.
3. Tsao, M.C., Walthall, B.J. and Ham, R.G. (1982) *J. Cell. Physiol.* **110,** 219.
4. Pintus, C., Ransom, J.H. and Evans, C.H. (1983) *J. Immunol. Meth.* **61,** 195.
5. Gilchrist, B.A., Marshall, W.C., Karassik, R.L., Weinstein, R. and Macaig, T. (1984) *J. Cell Physiol.* **120,** 377.
6. Jassal, D., Han, R.N., Caniggia, I., Post, M. and Tanswell, A.K. (1991) *In Vitro* **27A,** 625.

12. Hsu, T.C. and Kellogg, D.S. (1960) *Natl Cancer Inst.* **25**, 221.
13. Moore, G.E., Gerner, R.E. and Franklin, H.A. (1967) *J. Am. Med. Ass.* **199**, 519.
14. Parker, R.C. *et al.* (1957) *Ann. NY Acad. Sci.* **5**, 303.
15. Ham, R.G. (1965) *Proc. Natl Acad. Sci. USA* **53**, 288.
16. Bettger, W.J. (1981) *Proc. Natl Acad. Sci. USA* **78**, 5588.
17. McKeehan, W.L., Hamilton, W.G. and Ham, R.G. (1976) *Proc. Natl Acad. Sci. USA* **73**, 2023.
18. McKeehan, W.L., McKeehan, K.A., Hammond, S.L. and Ham, R.G. (1977) *In Vitro* **17**, 495.
19. Ringer, S. (1895) *J. Physiol.* **18**, 425.
20. Tyrode, M.V. (1910) *Arch. Intern. Pharmacodyn. Therap.* **22**, 205.
21. Earle, W.R., Schilling, E.L., Stark, T.H., Straus, N.P., Brown, M.F. and Shelton, E. (1943) *J. Natl Cancer Inst.* **4**, 165.
22. Hanks, J.H. and Wallace, R.E. (1949) *Proc. Soc. Exp. Biol. Med.* **71**, 196.
23. Gey, G.O. and Gey, M.K. (1936) *Am. J. Cancer.* **27**, 45.
24. Puck, T.T., Cieciura, S.J. and Fischer, H.W. (1957) *J. Exp. Med.* **106**, 145.
25. Puck, T.T., Cieciura, S.J. and Robinson, A. (1958) *J. Exp. Med.* **108**, 945.
26. Paul, J. (1975) *Cell and Tissue Culture*. Churchill Livingstone, Edinburgh.

7. Burgess, W.H., Mehlman, T., Freisel, R., Johnson, W.V. and Macaig, T. (1985) *J. Biol. Chem.* **260**, 11389.
8. Burgess, W.H., Mehlman, T., Marshall, D.R., Fraser, B.A. and Macaig, T. (1986) *Proc. Natl Acad. Sci. USA* **83**, 7216.
9. Carpenter, G. and Wahl, M.I. (1990) in *Handbook of Experimental Pharmacology: Peptide Growth Factors and Their Receptors* (M.B. Sporn and A.B. Roberts, eds). Springer-Verlag, Heidelberg.
10. Carpenter, G. (1987) *Ann. Rev. Biochem.* **56**, 881.
11. Shoemaker, C.B. and Mitsock, L.D. (1986) *Mol. Cell. Biol.* **6**, 849.
12. Kitamura, T., Tange, T., Terasawa, T., Chilba, S., Kuwaki, T., Miyogawa, K., Piao, Y.F., Miyazono, K., Urabe, A. and Takaku, F. (1989) *J. Cell. Physiol.* **140**, 323.
13. Carpenter, G. and Wahl, M.I. (1990) in *Handbook of Experimental Pharmacology: Peptide Growth Factors and Their Receptors* (M.B. Sporn and A.B. Roberts, eds). Springer-Verlag, Heidelberg.
14. Burgess, W.H. and Macaig, T. (1989) *Ann. Rev. Biochem.* **58**, 575.
15. Gospodarowicz, D. (1987) *Meth. Enzymol.* **147**, 106.
16. Nagata, S. (1990) in *Handbook of Experimental Pharmacology: Peptide Growth Factors and Their Receptors* (M.B. Sporn and A.B. Roberts, eds). Springer-Verlag, Heidelberg.

References

17. Shirafuji, N., Asano, S., Matsuda, S., Watan, K., Tokoku, F. and Nagata, S. (1989) *Exp. Hematol.* **17,** 116.

18. Metcalf, D. (1985) *Science* **229,** 16.

19. Tsuchiya, M., Asano, S., Kaziro, Y. and Nagata, S. (1986) *Proc. Natl Acad. Sci. USA* **83,** 7633.

20. Whetton, A.D. (1990) *Trends Pharmacol. Sci.* **11,** 285.

21. Rubin, J.S. *et al.* (1991) *Proc. Natl Acad. Sci. USA* **88,** 415.

22. Cooper, C.S. (1992) *Oncogene* **7,** 3.

23. Peska, S., Langer, J.A., Zoon, K.C. and Samuel, C.E. (1987) *Ann. Rev. Biochem.* **56,** 727.

24. Krim, M. (1980) *Blood* **55,** 875.

25. Palmer, H. and Libby, P. (1992) *Lab. Invest.* **66,** 715.

26. Krim, M. (1980) *Blood* **55,** 711.

27. Perussia, B., Dayton, E.T., Fanning, V., Thiagarajan, P., Hoxie, J. and Trinchieri, G. (1983) *J. Exp. Med.* **158,** 2058.

28. Gusella, G.L., Musso T., Bosco, M.C., Espinoza-Delgado, I., Matsushima, K. and Varesio, L. (1993) *J. Immunol.* **151,** 2725.

29. Froesch, E.R., Schmid, C., Schwandler, J. and Zapf, J. (1985) *Ann. Rev. Physiol.* **47,** 443.

30. Delafontaine, P. and Lou, H. (1993) *J. Biol. Chem.* **268,** 16866.

31. Elgin, R.G., Busby, W.H. and Clemmons, D.R. (1987) *Proc. Natl Acad. Sci. USA* **84,** 3254.

47. Goodwin, R.G., Lupton, S., Schmierer, A., Hjerrild, K.J., Jerzy, R., Clevenger, W., Gillis, S., Cosman, D. and Namen, A.E. (1989) *Proc. Natl Acad. Sci. USA* **86,** 302.

48. Billips, L.G., Petitte, D., Dorshkind, K., Narayanan, R., Chiu, C.P. and Landreth, K.S. (1992) *Blood* **79,** 1185.

49. Taub, D.D. and Oppenheim, J.J. (1993) *Cytokine* **5,** 175.

50. Strieter, P.M., Kunkel, S.L., Showell, S.J., Remick, D.G., Pham, S.H., Ward, P.A. and Marks, R.M. (1989) *Science* **243,** 1467.

51. Schroder, J.M., Sticherling, M., Henneicke, H.H., Preissner, W.C. and Christophers, E. (1990) *J. Immunol.* **144,** 2223.

52. Yang, Y.C., Ricciardi, S., Ciarletta, A., Calvetti, J., Kelleher, K. and Clark, S.C. (1989) *Blood* **74,** 1880.

53. Uyttenhove, C., Simpson, R.J. and Van Snick, J. (1988) *Proc. Natl Acad. Sci. USA* **85,** 6934.

54. Moore, K.W., Ogarra, A., Malefyt, R.D., Vieira, P. and Mossman, T.R. (1993) *Ann. Rev. Immunol.* **11,** 165.

55. Thompson-Snipes, L., Dhar, V., Bond, M.W., Mosman, T.R., Moore, K.W. and Rennick, D.M. (1991) *J. Exp. Med.* **173,** 507.

56. Paul, S.R. **et al.** (1990) *Proc. Natl Acad. Sci. USA* **87,** 7512.

57. Quesniaux, V.F.J., Clark, S.C., Turner, K. and Fagg, B. (1992) *Blood* **80,** 1218.

58. Scott, P. (1993) *Science* **260,** 496.

59. Gubler, U. *et al.* (1991) *Proc. Natl Acad. Sci. USA* **88,** 4143.

32. Smith, M.C., Cook, J.A., Furma, T.C. and Occolowitz, J.L. (1989) *J. Biol. Chem.* **264,** 9314.

33. Oppenheim, J.J., Kovacs, E.J., Matsushima, K. and Durum, S.K. (1986) *Immunol. Today* **7,** 45.

34. Dinarello, C.A. (1991) *Blood* **77,** 1627.

35. Mier, J.W. and Gallo, R.C. (1980) *Proc. Natl Acad. Sci. USA* **77,** 6134.

36. Brandhuber, B.J., Boone, T., Kenney, W.C. and McKay, D.B. (1987) *Science* **238,** 1707.

37. Smith, K.A. (1984) *Ann. Rev. Immunol.* **2,** 319.

38. Brazill, G.W., Haynes, M., Garland, J. and Dexter, T.M. (1983) *Biochem. J.* **210,** 747.

39. Morstyn, G. and Burgess, A.W. (1988) *Cancer Res.* **48,** 5624.

40. Miyajima, A., Mui, A.L.F., Ogorochi, T. and Sakamaki, K. (1993) *Blood* **82,** 1960.

41. Paul, W.E. (1991) *Blood* **77,** 1859.

42. Fernandez-Botran, Krammer, P.H., Diamantstein, T., Uhr, J.W. and Vitetta, E.S. (1986) *J. Exp. Med.* **164,** 580.

43. Takatsu, K. (1992) *Curr. Opin. Immunol.* **4,** 299.

44. Swain, S.L. (1985) *J. Immunol.* **134,** 3934.

45. Kishimoto, T., Akira, S. and Taga, T. (1992) *Science* **258,** 593.

46. Van Snick, J. (1990) *Ann. Rev. Immunol.* **8,** 253.

60. McKenzie, A.N.J. *et al.* (1993) *Proc. Natl Acad. Sci. USA* **90,** 3735.

61. Minty, A. *et al.* (1993) *Nature* **362,** 248.

62. Barnes, D. and Sato, G. (1980) *Anal. Biochem.* **102,** 255.

63. Aaronson, S.A., Bottaro, D.P., Miki, T., Ron, D., Finch, P.W., Fleming, T.P., Ahn, J., Taylor, W.G. and Rubin, J.S. (1991) *Ann. NY Acad. Sci.* **638,** 62.

64. Finch, P.W., Rubin, J.S., Miki, T., Ron, D. and Aaronson, S.A. (1991) *Science* **245,** 752.

65. Gearing, D.P., Gough, N.M., King, J.A., Hilton, D.J., Nicola, N.A., Simpson, R.J., Nice, E.C., Kelso, A. and Metcalf, D. (1987) *EMBO J.* **6,** 3995.

66. Williams, R.L., Hilton, D.J., Pease, S., Willson, T.A., Stewart, C.L., Gearing, D.P., Wagner, E.F., Metcalf, D., Nicola, N.A. and Gough, N.M. (1988) *Nature* **336,** 684.

67. Moreau, J., Donaldson, D.D., Bennett, F., Giannotti-Witek, J., Clark, S.C. and Wong, G.G. (1988) *Nature* **336,** 690.

68. Smith, A.G., Heath, J.K., Donaldson, D.D., Wong, G.G., Moreau, J., Stahl, M. and Rogers, D. (1988) *Nature* **336,** 688.

69. Pickart, L. (1981) *In Vitro* **17,** 459.

70. Pickart, L., Freedman, J.H., Loker, W.J., Peisach, J., Perkins, C.M., Stenkamp, R.E. and Weinstein, B. (1980) *Nature* **288,** 715.

71. Maquart, F.-X. *et al.* (1993) *J. Clin. Invest.* **92,** 2368.

References

72. Das, S.K., Stanley, E.R., Guilbert, L.J. and Forman, L.W. (1981) *Blood* **58**, 630.

73. Sherr, C.J. (1990) *Blood* **75**, 1.

74. Levi-Montalcini, R. (1987) *Science* **237**, 1154.

75. Stephani, U., Sutter, A. and Zimmermann, A. (1987) *J. Neurosci. Res.* **17**, 25.

76. Otten, U., Ehrhard, D. and Peck, R. (1989) *Proc. Natl Acad. Sci. USA* **86**, 10059.

77. Zarling, J.M., Shoyab, M., Marquardt, H., Hanson, M.B., Lioubin, M.N. and Todaro, G.J. (1986) *Proc. Natl Acad. Sci. USA* **83**, 9739.

78. Linsley, P.S., Boltonhanson, M., Horn, D., Malik, N., Kallestad, J.C., Ochs, V., Zarling J.M. and Shoyab, M. (1988) *J. Biol. Chem.* **264**, 4282.

79. Ross, R., Raines, E.W. and Bowenpope, D.F. (1986) *Cell* **46**, 155.

80. Soma, Y., Dvonch, V. and Grotendorst, G.R. (1992) *FASEB J.* **6**, 2996.

81. Carow, C.E., Hangoc, G., Cooper, S.H., Williams, D.E. and Broxmeyer, H.E. (1991) *Blood* **78**, 2216.

82. Martin, F.H. *et al.* (1990) *Cell* **63**, 203.

83. Massague, J. (1990) *J. Biol. Chem.* **265**, 21393.

84. Derynck, R. (1988) *Cell* **54**, 593.

85. Sporn, M.B. and Roberts, A.B. (1992) *J . Cell Biol.* **119**, 1017.

4. Hopps, H.E., Bernhaim, B.C., Nisalak, A., Tjio, J.H. and Smadel, J.E. (1963) *J. Immunol.* **91**, 416.

5. Gluzman, Y. (1981) *Cell* **23**, 175.

6. Jensen, F.C., Girardi, A.J., Gilden, R.V. and Koprowski, H. (1964) *Proc. Natl Acad. Sci. USA* **52**, 53.

7. Hull, R.N., Cherry, W.R. and Johnson, I.S. (1956) *Anat. Rec.* **124**, 490.

8. Kawakami, T.G., Huff, S.D., Buckley, P.M., Dungworth, P.D., Snyder, S.P. and Gilden, R.V. (1972) *Nature (New Biol.)* **235**, 170.

9. Robin, H., Hopkins, R.F., Ruscetti, F.W., Neubauer, R.H., Brown, R.L. and Kawakami, T.G. (1981) *J. Immunol.* **127**, 1852.

10. Yasumura, Y. and Kawakita, Y. (1963) *Nippon Rinsho* **21**, 1209.

11. Rhim, J.S. and Schell, K. (1967) *Proc. Soc. Exp. Biol. Med.* **125**, 602.

12. Rhim, J.S. and Schell, K. (1967) *Nature* **216**, 271.

13. Pattillo, R.A., Gey, G.O., Delfs, E. and Mattingly, R.F. (1968) *Science* **159**, 1467.

14. Foley, G.E., Lazarus, H., Farber, S., Uzman, B.G., Boone, J.B.A. and McCarthy, R.E. (1965) *Cancer* **18**, 522.

15. Adams, R.A. (1967) *Cancer Res.* **27**, 2479.

86. Massague, J. (1987) *Cell* **49**, 437.
87. Beutler, B. and Cerami, A. (1986) *Nature* **320**, 584.
88. Vassalli, P. (1992) *Ann. Rev. Immunol.* **10**, 411.

Chapter 4

1. Ramsey, W.S., Hertl, W., Nowlan, E.D. and Binkowski, N.J. (1984) *In Vitro* **20**, 802.
2. Van Wezel, A.L. (1967) *Nature* **216**, 64.
3. Lakhotia, S. and Papoutsakis, E.T. (1992) *Biotech. Bioeng.* **39**, 95.
4. Takagi, M. and Ueda, K. (1994) *J. Ferment. Bioeng.* **77**, 301.

Chapter 5

1. Hay, R., Caputo, J., Chen, T.R., Macy, M., McClintock, P. and Reid, Y. (eds) (1992) *American Tissue Type Culture Collection*, 7th Edn. ATCC, Rockville, MD.
2. Waymouth, C. (1974) *In Vitro* **10**, 97.
3. Miller G. and Lipman M. (1973) *Proc. Natl Acad. Sci. USA* **70**, 190.

16. Adams, R.A., Foley, G.E., Uzman, B.G., Lazarus, H. and Kleinman, L. (1967) *Cancer Res.* **27**, 772.
17. Chang, R.S.M. (1954) *Proc. Soc. Exp. Biol. Med.* **87**, 440.
18. Klein, E., Klein, G., Nadkarni, J.S., Nadkarni, J.J., Wigzell, H. and Clifford, P. (1968) *Cancer Res.* **28**, 1300.
19. Peterson, W.D., Stulberg, C.S. and Simpson, W.E. (1971) *Proc. Soc. Exp. Biol. Med.* **136**, 1187.
20. Epstein, M.A. and Barr, Y.M. (1964) *Lancet* **1**, 252.
21. Gey G.O., Coffman W.D. and Kubicek M.T. (1952) *Cancer Res.* **12**, 264.
22. Jones, H.W., McKusick V.A., Harper, P.S. and Wuu, K. (1971) *Obstet. Gynecol.* **38**, 945.
23. Puck, T.T., Marcus, P.I. and Cieciura, S.J. (1956) *J. Exp. Med.* **103**, 273.
24. Toolan, H.W. (1954) *Cancer Res.* **14**, 660.
25. Collins, S.J., Gallo, R.C. and Gallagher, R.E. (1977) *Nature* **270**, 347.
26. Wang, A.M., Creasy, A.A., Ladner, M.B., Lin, L.S., Strickler, J., Van Arsdell, J.N., Yamamoto, R. and Mark, D.F. (1985) *Science* **228**, 149.
27. Fahey, J.L., Buell, D.N. and Sox, H.C. (1972) *Ann. NY Acad. Sci.* **190**, 221.
28. van Boxel, J.A. and Buell, D.N. (1974) *Nature* **251**, 443.

29. van Obberghen, E., de Meyts, P. and Roth, J. (1976) *J. Biol. Chem.* **251**, 6844.

30. Lozzio, C.B. and Lozzio, B.B. (1975) *Blood* **45**, 321.

31. Eagle, H. (1955) *Proc. Soc. Exp. Biol. Med.* **89**, 362.

32. Koeffler, H.P. and Golde, D.W. (1978) *Science* **200**, 1153.

33. Soule, H.D. *et al.* (1973) *J. Natl Cancer Inst.* **51**, 1409.

34. Minowada, J., Ohnuma, T. and Moore, G.E. (1972) *J. Natl Cancer. Inst.* **49**, 891.

35. Klein, G., Dombos, L. and Gothoskar, B. (1972) *Int. J. Cancer* **10**, 44.

36. Pulvertaft, R.J.V. (1964) *Lancet* **1**, 238.

37. Huang, C.C. and Moore, G. (1969) *J. Natl Cancer Inst.* **43**, 1119.

38. Moore, G.E. and Sandberg, A.A. (1964) *Cancer* **17**, 170.

39. Matsuoka, Y., Moore, G.E., Yagi, Y. and Pressman, D. (1967) *Proc. Soc. Exp. Biol. Med.* **125**, 1246.

40. Bubenik, J. *et al.* (1970) *Int. J. Cancer* **5**, 310.

41. Hayflick, L. (1961) *Exp. Cell Res.* **23**, 14.

42. Macpherson, I. and Stoker, M. (1962) *Virology* **16**, 147.

43. Macpherson, I. (1963) *J. Natl Cancer Inst.* **30**, 795.

44. Ham, R.G. (1965) *Proc. Natl Acad. Sci. USA* **53**, 288.

45. Hsu, T.C. and Zenzes, M.T. (1964) *J. Natl Cancer Inst.* **32**, 857.

46. Aaronson, S.A. and Todero, G.J. (1968) *J. Cell Physiol.* **72**, 141.

60. Dexter, T.M., Garland, J., Scott, E., Scolnick, E. and Metcalf, D. (1980) *J. Exp. Med.* **152**, 1036.

61. Spooncer, E., Heyworth, C.M., Dunn, A. and Dexter, T.M. (1986) *Differentiation* **31**, 111.

62. Markowitz, D., Goff, S. and Bank, A. (1988) *J. Virol.* **62**, 1120.

63. Levine, L. *et al.* (1972) *Biochem. Biophys. Res. Commun.* **47**, 888.

64. Sanford, K.K., Earle, W.R. and Likely, G.D. (1948) *J. Natl Cancer Inst.* **9**, 229.

65. Moore, G.E., Sandberg, A.A. and Ulrich, K. (1966) *J. Natl Cancer Inst.* **36**, 405.

66. Bertram, J.S. and Janik, P. (1980) *Cancer Lett.* **11**, 63.

67. Potter, M. and Robertson, C.L. (1960) *J. Natl Cancer Inst.* **25**, 847.

68. Yosida, T.H., Imai, H.T. and Potter, M. (1968) *J. Natl Cancer Inst.* **41**, 1083.

69. Laskov, R. and Scharff, M.D. (1970) *J. Exp. Med.* **131**, 515.

70. Buonassisi, V., Sato, G. and Cohen, A.I. (1962) *Proc. Natl Acad. Sci. USA* **48**, 1184.

71. Sandford, K.K., Merwin, R.M., Hobbs, G.L., Young, J.M. and Earle, W.R. (1959) *J. Natl Cancer Inst.* **23**, 1035.

72. Klebe, R.J. and Ruddle, F.H. (1969) *J. Cell Biol.* **43**, 69A.

73. Jainchill, J.L., Aaronson, S.A. and Todaro, G.J. (1969) *J. Virol.* **4**, 549.

47. Todaro, G.J. and Green, H. (1963) *J. Cell Biol.* **17,** 299.
48. Markowitz, D., Goff, S. and Bank, A. (1986) *Virology* **167,** 400.
49. Hart, I.R. (1979) *Am. J. Path.* **97,** 587.
50. Fidler, I.J. (1973) *Nature (New Biol.)* **242,** 148.
51. Ralph, P. (1973) *J. Immunol.* **110,** 1470.
52. Goldsby, R.A., Osborne, B.A., Simpson, E. and Herzenberg, L.A. (1977) *Nature* **267,** 707.
53. Reznikoff, C., Brankow, D. and Heidelberger, C. (1973) *Cancer Res.* **33,** 3231.
54. Lassar, A.B., Paterson, B.M. and Weintraub, H. (1986) *Cell* **47,** 649.
55. Yasumura, Y., Tashjian, A.H. and Sato, G.H. (1966) *Science* **154,** 1186.
56. Boone, C., Sasaki, M. and McKee, W. (1965) *J. Natl Cancer Inst.* **34,** 725.
57. Doetschman, T.C., Eistetter, H., Katz, M., Schmidt., W. and Kemler, R. (1985) *Embryol. Exp. Morphol.* **87,** 27.
58. Williams, D.L., Hilton, D.J., Pease, S., Willson, T.A., Stewart, C.L., Gearing, D.P., Wagner, E.F., Metcalf, D., Nicola, N.A. and Gough, N.M. (1988) *Nature* **336,** 684.
59. Gossler, A., Doetschman, T., Korn, R., Serfling, E. and Kemler, R. (1986) *Proc. Natl Acad. Sci. USA* **83,** 9065.

74. Kohler, G. and Milstein, C. (1976) *Eur. J. Immunol.* **6,** 511.
75. Kohler, G. and Milstein, C. (1975) *Nature* **256,** 495.
76. Kearney, J., Radbruch, A., Liesegang, B. and Rajewsky, K. (1979) *J. Immunol.* **123,** 1548.
77. Dawe, C.J. and Potter, M. (1957) *Am. J. Pathol.* **33,** 603.
78. Plaut, M., Lichtenstein, L.M., Gillespie, E. and Henry, C.S. (1973) *J. Immunol.* **111,** 389.
79. Miller, A.D. and Buttimore C. (1986) *Mol. Cell. Biol.* **6,** 2895.
80. Schulman, M., Wilde, C.D. and Kohler, G. (1978) *Nature* **276,** 269.
81. Yasumura, Y., Buonassisi, V. and Sato, G. (1966) *Cancer Res.* **26,** 529.
82. Kiessling, R., Klein, E. and Wigzell, H. (1975) *Eur. J. Immunol.* **5,** 112.
83. Benda, P., Lightbody, J., Sato, G., Levine, L. and Sweet, W. (1968) *Science* **161,** 370.
84. Ambesi-Impiombato, F.S., Parks, L.A.M. and Coon, H.G. (1980) *Proc. Natl Acad. Sci. USA* **77,** 3455.
85. Valente, W.A., Vitti, P., Kohn, L.D., Brandi, M.L., Rotella, C.M., Roberto, T., Tramontano, D., Aloji, S.M. and Ambesi-Ipiombato, F.S. (1983) *Endocrinology* **112,** 71.
86. Yasumura, Y., Tashjian, A.H. and Sato, G. (1966) *Science* **154,** 1186.

133

References

87. Tashjian, A.H., Yasumura, Y., Levine, L., Sato, G.H. and Parker, M.L. (1968) *Endocrinology* **82**, 342.

88. McCoy, T.A., Maxwell, M. and Kruse, P.F. (1959) *Cancer Res.* **19**, 591.

89. Richardson, U.I., Tashjian, A.H. and Levine, L. (1969) *J. Cell Biol.* **40**, 236.

90. De Larco, J.E. and Todaro, G.J. (1978) *J. Cell. Physiol.* **94**, 335.

91. Galfre, G., Milstein, C. and Wright, B. (1979) *Nature* **277**, 131.

92. Kilmartin, J.V., Wright, B. and Milstein, C. (1982) *J. Cell Biol.* **93**, 576.

93. Crandell, R.A., Fabricant, C.G. and Nelson-Rees, W.A. (1973) *In Vitro* **9**, 176.

94. Armstrong, J.A., Porterfield, J.S. and de Madrid, A.T. (1971) *J. Gen. Virol.* **10**, 195.

95. Hull, R.N., Dwyer, A.C., Cherry, W.R. and Tritch, O.J. (1965) *Proc. Soc. Exp. Biol. Med.* **118**, 1054.

96. Madin, S.H. and Darby, N.B. (1958) *Proc. Soc. Exp. Biol. Med.* **98**, 574.

97. Rindler, M.J., Chuman, L.M., Shaffer, L. and Saier, M.H. (1979) *J. Cell Biol.* **81**, 635.

98. Montesano, R., Schaller, G. and Orci, L. (1991) *Cell* **66**, 697.

99. Peebles, P.T., Gerwin, B.I., Papageorge, A.G. and Smith, S.G. (1975) *Virology* **67**, 344.

115. Rous, P. and Jones, F.S. (1916) *J. Exp. Med.* **23**, 549.

116. Peoples, G.E., Goedegebuure, P.S., Andrews, J.V.R., Schoof, D.D. and Eberlein, T.J. (1993) *J. Immunol.* **151**, 5481.

117. Ives, H.E., Schultz, G.S., Galardy, R.E. and Jamieson, J.D. (1978) *J. Exp. Med.* **148**, 1400.

118. Reubi, J.C., Horisberger, U., Long, W., Koper, J.W., Brookman, R. and Lamberts, S.W.J. (1989) *Am. J. Pathol.* **134**, 337.

119. Abken, H., Jungfer, H., Albert, W.H.W. and Willecke, K. (1986) *J. Cell Biol.* **103**, 795.

120. Hemstreet, G.P., Enoch, P.G. and Pretlow, T.G. (1980) *Cancer Res.* **40**, 1043.

121. Rinaldin, K.M. (1959) *Exp. Cell Res.* **16**, 477.

122. Bashor, M.M. (1979) *Meth. Enzymol.* **58**, 119.

123. Halaban, R. and Alfano, F.D. (1984) In Vitro *Cell. Dev. Biol.* **20**, 447.

124. Rheinwald, J.G. and Green, H. (1975) *Cell* **6**, 331.

125. Henning, M., Michael, D., Cheng, C., Steinert, P. and Holbrook, K. (1980) *Cell* **19**, 245.

126. Willie, J.J., Pittelkow, M.R., Shipley, G.D. and Scott, R.E. (1984) *J. Cell. Physiol.* **121**, 31.

127. Gilbert, S.F. and Migeon, B.R. (1975) *Cell* **5**, 11.

128. Sordillo, L.M., Oliver, S.P. and Akers, R.M. (1988) *Cell Biol. Int. Rep.* **12**, 355.

100. Wurster, D.H. and Benhirschke, K. (1970) *Science* **168,** 1364.

101. Phillips, C.A., Melnick, J.L. and Burkardt, M. (1966) *Proc. Soc. Exp. Biol. Med.* **122,** 783.

102. Nichols, W.W., Murphy, D.G., Cristofalo, V.J., Toji, L.H., Greene, A.E. and Dwight, S.A. (1977) *Science* **196,** 60.

103. Jacobs, J.P., Jones, C.M. and Baille, J.P. (1970) *Nature* **227,** 168.

104. Jacobs, J.P., Garrett, A.J. and Merton, R. (1979) *J. Biol. Stand.* **7,** 113.

105. Hayflick, L. and Moorhead, P.S. (1961) *Exp. Cell Res.* **25,** 585.

106. Schaeffer, W.I. and Waymouth, C. (1976) in *Cell Biology I, Biological Handbook* (P.L. Altman and D.D. Katz, eds), p. 46. FASEB, Bethesda, MD.

107. Matsumura, T. *et al.* (1975) *Jap. J. Exp. Res.* **45,** 377.

108. Steinberg, M.S. (1963) *Exp. Cell Res.* **30,** 257.

109. Phillips, H.J. (1972) *In Vitro* **8,** 101.

110. Amsterdam, A. and Jamieson, J.D. (1972) *Proc. Natl Acad. Sci. USA* **69,** 3028.

111. Gibsen, T.L., Bolognese, A., Maddrell, C., Steffek, A.J. and Forbes, D.P. (1989) *J. Craniofacial Gene Dev. Biol.* **9,** 349.

112. Bashor, M.M. (1979) *Methods Enzymol.* **58,** 119.

113. Weinstein, D. (1966) *Exp. Cell Res.* **43,** 234.

114. Frazier *et al.* (1975) *Lab. Invest.* **33,** 231.

129. Mason, E.A., Atkin, S.L. and White, M.C. (1993) In Vitro *Cell. Dev. Biol.* **29A,** 912.

130. Logsdon, C.D and Williams, J.A. (1986) *Am. J. Physiol.* **250,** 440.

131. Taylor-Papadimitriou, J. (1992) in *Culture of Epithelial Cells* (I.R. Freshney, ed.), p. 107. Wiley-Liss, New York.

132. Kao, W.W. and Prockop, D.J. (1977) *Nature* **266,** 63.

133. Stoner, G.D. and Klaunig, J.E. (1983) in *Cell Separation: Methods and Selected Applications* (T.P. Pretlow and T.F. Pretlow, eds), Vol. 2. Academic Press, New York.

134. Paraskeva, C., Buckle, B.G. and Thorpe, P.E. (1985) *Br. J. Cancer* **51,** 131.

135. Singer, K.H., Scoarce, R.M. and Tuck, D.T. (1989) *J. Invest. Dermatol.* **92,** 166.

136. Linge, C., Green, M.R. and Brooks, R.F. (1989) *Exp. Cell Res.* **185,** 519.

137. Kirkland, S.C. and Bailey, I.G. (1986) *Br. J. Cancer* **53,** 779.

Chapter 6

1. Hudson, L. and Hay, F.C. (1991) *Practical Immunology,* p. 92. Blackwell Scientific Publications, Oxford.

2. Fechheimer, M., Boylan, J.F., Parker, S., Siskens, J.E., Patel, G.L. and Zimmer, S.G. (1987) *Proc. Natl Acad. Sci. USA* **84,** 8463.

3. Testa, N.G. and Molineux, G. (1993) *Haemopoiesis: A Practical Approach*. IRL Press, Oxford.

4. Clive, D., McCuen, R., Spector, J.F.S., Piper, C. and Malvournin, K.H. (1983) *Mutat. Res.* **115**, 225.

5. Singer, B. and Kusmierek, J.T. (1982) *Ann. Rev. Biochem.* **51**, 655.

6. Hooper, M.L. (1985) in *Mammalian Cell Genetics,* p. 47. Wiley, New York.

7. King, W., Patel, P., Lobel, L.I., Goff, S.P. and Chi Nguyen-Huu, M. (1988) *Science* **228**, 554.

8. Stocking, C., Loliger, C., Kawai, M., Suciu, S., Gough, N. and Ostertag, W. (1988) *Cell* **53**, 869.

9. Lobel, L.I., Patel, M., King, W., Nguyen-Huu, M.C. and Goff, S.P. (1984) *Science* **228**, 329.

10. Gaken, J. and Farzaneh, F. (1991) in *Methods in Molecular Biology* (M. Collins, ed.), Vol. 8, p. 111. Humana Press, Totoua, NJ.

11. Capecchi, M.R. (1989) *Science* **244**, 1288.

12. Mansour, S.L., Thomas, K.R. and Capecchi, M.R. (1988) *Nature* **336**, 348.

13. Kohler, G. and Milstein, C. (1975) *Nature* **256**, 495.

14. Johnstone, A. and Thorpe, R. (1987) *Immunochemistry in Practice,* p. 31. Blackwell Scientific Publications, Oxford.

15. Hudson, L. and Hay, F.C. (1991) *Practical Immunology,* p. 14. Blackwell Scientific Publications, Oxford.

25. Hancock, J.F. (1991) in *Methods in Molecular Biology* (M. Collins, ed.), Vol. 8, p. 153. Humana Press, Totoua, NJ.

26. Itani, T., Ariga, H., Yamaguchi, N., Tadakuma, T., and Yasuda, T. (1987) *Gene* **56**, 267.

27. Potter, H.L., Weir, U., and Leder, P. (1984) *Natl Acad. Sci. USA* **81**, 7161.

28. Neumann, E., Schaefner-Ridder, M., Wang, Y. and Hofschneider, P.H. (1982) *EMBO J.* **1**, 841.

29. Klein, T.M., Wolf, E.D., Wu, R. and Sanford, J.C. (1987) *Nature* **327**, 70.

30. Sanford, J.C., Smith, F.D. and Russell, J.A. (1993) *Meth. Enzymol.* **271**, 483.

31. Yokoyama, K. and Imamoto, F. (1987) *Proc. Natl Acad. Sci. USA* **84**, 7363.

32. Capecchi, M.R. (1980) *Cell* **22**, 479.

33. Markowitz, D., Goff, S. and Bank, A. (1988) *Virology* **167**, 400.

34. Levrero, M., Barban, V., Manteca, S., Ballary, A., Balsamo, C., Avanitaggiat, M.L., Natoli, G., Skellekens, H., Tiollas, P. and Perricaudet, M. (1991) *Gene* **101**, 195.

35. Wagner, E., Zatlouka, K., Cotten, M., Kirlappos, H., Mechtler, K., Curiel, D. and Birnstiel, M. (1992) *Proc. Natl Acad. Sci. USA* **87**, 6099.

16. Harlow, E. and Lane, D. (1988) *Antibodies: a Laboratory Manual.* Cold Spring Harbor Scientific Publications, New York.
17. Zanella, I., Verardi, R., Negrini, R., Poiesi, C., Ghielmi, S. and Albertini, A. (1992) *J. Immunol. Meth.* **156,** 205.
18. Kozbor, D., Roder, J.C., Sierzega, M.C., Cole, S.P.C. and Croce, C.M. (1986) *Meth. Enzymol.* **121,** 120.
19. Morgan, S.J. and Darling, D.C. (1993) *Animal Cell Culture.* BIOS Scientific Publishers, Oxford.
20. Pakkanen, R. and Neutra, M. (1994) *J. Immunol. Meth.* **169,** 63.
21. Graham, F.L. and Van Der Eb, A.J. (1973) *Virology* **52,**456.
22. Chu, G. and Sharp, P.A. (1981) *Gene* **13,** 197.
23. Chen, C.A. and Okayama, H. (1988) *Biotechniques* **6,** 932.
24. Somapayrac, L.M. and Danna, K.J. (1981) *Proc. Natl Acad. Sci. USA* **78,** 7575.

36. Cotten, M., Wagner, E., Zatlouka, K., Phillips, S., Curiel, D. and Birnstiel, M. (1992) *Proc. Natl Acad. Sci. USA* **87,** 6094.
37. Blochlinger, K. and Diggleman, H. (1984) *Mol. Cell. Biol.* **4,** 2929.
38. Southern, P.J. and Berg, P. (1982) *J. Mol. Appl. Gen.* **1,** 327.
39. de la Luna, S. and Ortin, J. (1982) *Meth. Enzymol.* **216,** 376.
40. Hartmann, S.C. and Mulligan, R.C. (1988) *Proc. Natl Acad. Sci. USA* **85,** 8047.
41. Mulligan, R.C. and Berg, P. (1981) *Proc. Natl Acad. Sci. USA* **78,** 2072.
42. Semon, D., Movva, N.R. and Smith, T.F. (1987) *Plasmid* **17,** 46.
43. Mulsant, P., Gatignol, A., Dalens, M. and Tiraby, G. (1988) *Som. Cell Mol. Genet.* **14,** 243.
44. Crouse, G.F., McEwan, R.N. and Pearson, M.L. (1983) *Mol. Cell. Biol.* **3,** 257.
45. Littlefield, J.W. (1964) *Science* **145,** 709.

137

APPENDIX

Safety

Safety is the most important part of any laboratory work. We will point out the most common day-to-day problems and dilemmas that can/will arise. We do not seek to replace recommendations issued either locally in your institute, or nationally.

In order to satisfy safety regulations for work carried out in the United Kingdom you or your supervisor must consult your local safety officer and/or the Health and Safety Executive. Useful publications on the categorization of this work include:

Advisory Committee on Dangerous Pathogens: Categorization of Pathogens According to Hazard and Categories of Containment (2nd Edn) (1990).

3. Fresh human primary tissue where the source is known to be safe (i.e. when the patient has been tested for infectious agents).
4. The rest of the cell culture repertoire involving growth of human and animal cell lines should still be treated with respect, but has less obvious hazards.

You should also be aware that modification of cells has the potential to make them more hazardous than they were previously.

The art of cell culture involves *protecting yourself* from the cells whilst at the same time protecting the cells from contamination. Over and above normal laboratory safety rules, you should:

1. Wear a protective (dedicated for cell culture) laboratory coat.

Both are available on request from The Health and Safety Executive, Library and Information Services, Baynards House, 1 Chepstow Place, Westbourne Grove, London, W2 4TF, UK.

For work in other countries, consult your local Health and Safety Office.

1 Handling cells

Cell culture is very varied, but activities can be ranked in order of potential hazard. Top of the list are:

1. Human and primate cells harboring known pathogenic organisms. These (for instance HIV) must not be worked on without specially designed containment laboratories. These are not dealt with here.
2. Fresh human primary tissue where little is known about the source (for instance fresh human lymphocytes can contain infectious agents such as HIV and/or hepatitis virus).

2. Wear gloves.
3. Beware of aerosol production; wear a mask if necessary.
4. Work in the appropriate level of containment – most cell culture can take place in class II cabinets, this protects both the operator and the culture. Hazards such as HIV and hepatitis require more stringent control.
5. Wash hands both before and after work with cells.
6. Decontaminate waste before disposal, i.e. liquid waste with bleach: 10 000 p.p.m. chlorine for 12 h.
7. All solid waste (including sharps) should be autoclaved in a validated autoclave before leaving the suite of laboratories.

2 Chemical hazards

Chemical hazards in cell culture should be treated in exactly the same manner as chemical hazards in the rest of the laboratory. *All* procedures concerning any chemical or biological work should be subject to COSHH (the Control of Substances Hazardous to Health) assessment.

INDEX

Index

Index

ESSENTIAL DATA SERIES

All researchers need rapid access to data on a daily basis. The *Essential Data* series provides this core information in convenient pocket-sized books. For each title, the data have been carefully chosen, checked and organized by an expert in the subject area. *Essential Data* books therefore provide the information that researchers need in the form in which they need it.

Centrifugation
D. Rickwood, T.C. Ford & J. Steensgaard
0 471 94271 5, March 1994, £12.95/$19.95

Gel Electrophoresis
D. Patel
0 471 94306 1, March 1994, £12.95/$19.95

Light Microscopy
C. Rubbi
0 471 94270 7, April 1994, £12.95/$19.95

Vectors
P. Gacesa & D. Ramji
0 471 94841 1, September 1994, £12.95/$19.95

Human Cytogenetics
D. Rooney & B. Czepulkowski (Eds)
0 471 95076 9, October 1994, £12.95/$19.95

Animal Cells: culture and media
D.C. Darling & S.J. Morgan
0 471 94300 2, October 1994, £12.95/$19.95

Nucleic Acid Hybridization
P. Gilmartin
0 471 95084 X, due 1994, £12.95/$19.95

Enzymes in Molecular Biology
C.J. McDonald (Ed.)
0 471 94842 X, due 1995, £12.95/$19.95

ORDER FORM

Please send me:

Qty	Title	Price/copy	Total
....	....	....	....
....	....	....	....
....	....	....	....

All prices correct at time of going to press but subject to change. Your order will be processed without delay, please allow 21 days for delivery. We will refund your payment without question if you return any unwanted book to us in re-saleable condition within 30 days. All books are available from your bookseller.

Method of payment

☐ Payment £/$_____ enclosed (payable to John Wiley & Sons Ltd).
  Orders for one book only – please add £3.00/$4.95 to cover postage and handling. Two or more books postage FREE.
☐ Purchase order enclosed
☐ Please send me an invoice
  (£3.00/$4.95 will be added to cover postage and handling)
☐ Please charge my credit card account
  ☐ American Express   ☐ Diners Club
  ☐ Visa   ☐ Mastercard

Card no. |__|__|__|__|__|__|__|__| Expiry: |__|__|

Signature: _____

Telephone our Customer Services Dept with your cash or credit card order on 01243 829121 or dial FREE on 0800 243407 (UK only)

Send my order to:

Name (PLEASE PRINT) _____

Position: _____

Address: _____

_____

_____

_____

Telephone: _____

Signature: _____ Date: _____

Return to: Rebecca Harfield, John Wiley & Sons Ltd, Baffins Lane, Chichester, West Sussex PO19 1UD, UK. Telefax: (01243) 539132
  or: Wiley-Liss, 605 Third Avenue, New York, NY 10158-0012, USA. Telefax: (212) 850-8888

☐ If you do not wish to receive mailings from other companies please tick this box or notify the Marketing Services Department at John Wiley & Sons Ltd.

Ⓦ **WILEY**